# 63rd International Scientific Conference on Experimental Stress Analysis (EAN 2025)



Full Proceedings

Klaster Hradiste nad Jizerou, Czech Republic
27-29 May 2025

ISBN: 979-8-3313-3172-6

**Printed from e-media with permission by:**

Curran Associates, Inc.
57 Morehouse Lane
Red Hook, NY 12571

**Some format issues inherent in the e-media version may also appear in this print version.**



Technical University of Liberec
Studentská 1402/2
461 17 Liberec 1
Czech Republic

Phone: +420 485 351 111

international@tul.cz

**Additional copies of this publication are available from:**

Curran Associates, Inc.
57 Morehouse Lane
Red Hook, NY 12571 USA
Phone: 845-758-0400
Fax: 845-758-2633
Email: curran@proceedings.com
Web: www.proceedings.com

# EXPERIMENTAL STRESS ANALYSIS 2025

63rd International Conference
27th – 29th May 2025, Klášter Hradiště nad Jizerou

## EAN 2025 - 63rd Conference on Experimental Stress Analysis, Proceedings of Full Papers

EDITORS: Michal Sivčák
Iva Petríková

TECHNICAL UNIVERSITY OF LIBEREC

AND

CZECH SOCIETY FOR MECHANICS

Book title EAN 2025 - 63rd Conference on Experimental Stress Analysis Proceedings of Full Papers

Editors Ing. Michal Sivčák, Ph.D.

prof. Ing. Iva Petríková, Ph.D.

# Main topics of the conference

- Development of experimental methods in mechanics
- Development of experimental methods in biomechanics
- New methods and applications of deformation and stress analysis in mechanical structures
- New methods and applications of deformation and stress analysis in civil and other structures
- Experiment as a tool for the verification of analytical and numerical methods
- Experimental research and prediction of strength, life and operational reliability of structures and devices
- Monitoring of operational loadings and operational states of structures and devices
- Mechanics of advanced material, composites and 3D printed material
- Methods and means of teaching experimental methods

## Scientific Committee

Ing. Karel Doubrava, Ph.D. – *Czech Technical University in Prague, CZE*

prof. Ing. Peter Frankovský, Ph.D. – *Technical University of Košice, SVK*

prof. Ing. Nikolaj Ganev, CSc. – *Czech Technical University in Prague, CZE*

prof. Ing. Radim Halama, Ph.D. – *Technical University of Ostrava, CZE*

prof. Ing. Stanislav Holý, CSc. – *Czech Technical University in Prague, CZE*

Ing. Lubomír Houfek, Ph.D. – *Brno University of Technology, CZE*

prof. RNDr. Miroslav Hrabovský, DrSc. – *Palacký University in Olomouc, CZE*

doc. Ing. Róbert Huňady, PhD. – *Technical University of Košice, SVK*

prof. Ing. Josef Jíra, CSc. – *Czech Technical University in Prague, CZE*

Ing. Kamil Kolařík, Ph.D. – *Czech Technical University in Prague, CZE*

Ing. Jan Krystek, Ph.D. – *University of West Bohemia, CZE*

doc. Ing. Tomáš Návrat, Ph.D. – *Brno University of Technology, CZE*

doc. Ing. Pavel Padevět, Ph.D. – *Czech Technical University in Prague, CZE*

Prof. Ing. Iva Petríková, Ph.D. – *Technical University of Liberec, CZE*

Dr. Ing. Roman Růžek – *Czech Aerospace Research Centre, CZE*

prof. Ing. Milan Růžička, CSc. – *Czech Technical University in Prague, CZE*

Ing. Jaroslav Václavík – *Research and Testing Institute Plzeň, CZE*

## Organizing Committee

Iva Petríková

Jarmila Čálková

Michal Sivčák

**We would like to express our sincere gratitude to our supervisors**

DEWESoft®

HBP
HBM měřicí technika

LABTECH

# Table of Contents

# Effective Mechanical Characteristics of 3D Printed Structures with Gradient Elastic Characterics

V. Afanaseva[1], A. Jíra[1,*]

[1] *Department of Mechanics, Faculty of Civil Engineering, Czech Technical University in Prague, Thákurova, Prague, Czech Republic*

* *jira@fsv.cvut.cz*

**Abstract:** This study explores the application of topologically optimized gyroid lattice structures made from titanium alloy Ti6Al4V, fabricated via selective laser melting (SLM), for use in implantology. The research focuses on the mechanical properties of porous structures with varying morphologies, specifically those exhibiting gradients in either wall thickness or cell size. The primary objective was to assess how these geometric variations influence the global elastic modulus, stress distribution, and potential for osseointegration.

**Keywords:** 3D printing; surface etching; titanium alloy; porosity; mechanical testing.

## 1 Introduction

Additive manufacturing of titanium alloys represents a rapidly advancing and highly promising direction in the field of implantology, where key requirements include high mechanical strength, biocompatibility, and the capacity for long-term integration with host bone tissue—commonly referred to as osseointegration [1]. Among the cutting-edge approaches to optimizing implant design is the utilization of topologically optimized porous architectures. In this context, triply periodic minimal surface (TPMS) structures, particularly gyroid geometries, have gained significant attention due to their favorable mechanical and biological properties. These geometries support the fabrication of both shell-type and strut-based architectures, enabling a high degree of design flexibility and performance optimization [2].

Gyroid structures are frequently employed as functional layers in intraosseous implants, where they serve as osseointegrative scaffolds that facilitate the natural ingrowth of bone tissue into the implant body [3]. These structures are commonly manufactured using Selective Laser Melting (SLM), an additive manufacturing technique that allows for precise control of geometry, low internal porosity, and the realization of complex design features directly from digital models [4]. SLM processing of titanium alloys such as Ti6Al4V has proven particularly advantageous due to its ability to retain mechanical integrity while offering high design resolution.

Porosity plays a crucial role in determining both the mechanical and biological behavior of gyroid structures. A high degree of open porosity enhances nutrient transport and provides an interconnected environment conducive to bone cell migration and vascularization, which are critical to successful osseointegration. However, increased porosity is often associated with a reduction in mechanical stiffness and strength, which poses a challenge in load-bearing applications [5, 6]. To address this trade-off, graded porosity designs have been proposed and investigated. These allow for the strategic distribution of dense and porous regions—denser zones near the bone–implant interface improve load transmission and mechanical anchorage, while more porous internal regions promote biological integration and mimic the anisotropic nature of trabecular bone [7].

Such architecturally tailored implants thus combine the mechanical stability necessary for clinical performance with the biological functionality required for long-term success, bringing them closer to the structural and functional complexity of native bone tissue.

This study focuses on the experimental evaluation of two fundamental morphological variants of gyroid structures fabricated via SLM using the Ti6Al4V alloy [8]. The first variant exhibits a gradient in wall thickness, while the second variant incorporates a gradient in unit cell size. These structural variations were selected to investigate the influence of geometric gradation on mechanical response and potential for biological performance. Cylindrical specimens, dimensionally representative of clinical dental implants, were produced and subjected to uniaxial compression testing to assess their load-bearing capacity and deformation behavior.

It is hypothesized that geometrical variation within the gyroid architecture enhances stress distribution, mitigates the risk of mechanical failure at critical interfaces (e.g., bone–implant junction or homogeneous–porous transition zones), and simultaneously improves cell adhesion and proliferation through increased surface area and enhanced permeability for nutrient exchange [9, 10]. This study contributes to the broader development of functionally graded biomaterials and personalized implants tailored to patient-specific anatomical and biomechanical requirements [11]. The primary objective was to analyze the experimental results and determine the global elastic modulus of the specimens, a key parameter for evaluating the mechanical performance and stiffness compatibility of the porous structures [12].

## 2 Methodology

A total of 40 test specimens were prepared for mechanical testing. The gyroid structure was selected as the unit cell of the porous structure, which can be defined by Eq. (1). The basic mathematical formulation is based on the following equation:

$$F(x, y, z) = t = \sin(\bar{x}) \cdot \cos(\bar{y}) + \sin(\bar{y}) \cdot \cos(\bar{z}) + \sin(\bar{z}) \cdot \cos(\bar{x}) \tag{1}$$

Here, $F$ defines the surface shape based on a three-dimensional function, and $t$ is a constant that directly influences the mean curvature of the gyroid surface. An approximation of the minimal gyroid surface is achieved when the constant $t = 0$. Additional parameters such as $\bar{x}$, $\bar{y}$, and $\bar{z}$ represent modified spatial coordinates defined as $\bar{x} = 2\pi x/a$, $\bar{y} = 2\pi y/a$, and $\bar{z} = 2\pi z/a$. In these expressions, $x$, $y$, and $z$ are the actual spatial coordinates, and the parameter $a$ determines the size of the cube into which the gyroid structure is inscribed. A gyroid cell defined in this way is referred to as a Single Gyroid ($F = F_{sg}$), and its applications are wide-ranging [8, 13].

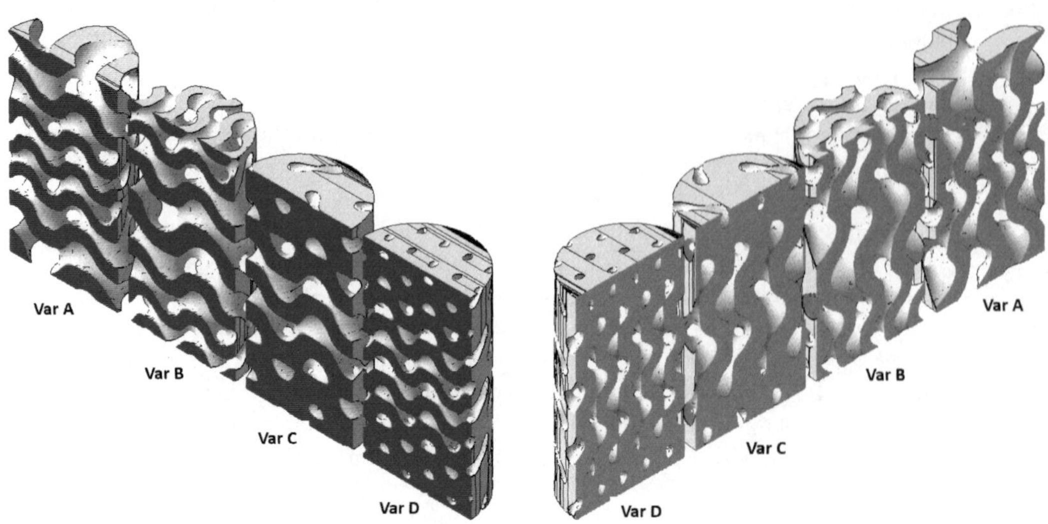


Fig. 1: Cross-sections of the test specimens showing different gradation patterns.

For the design of the test specimens, the general gyroid Eq. (1) was used, with a modified wall character to enable a smooth transition in the morphology of the gyroid cell. Two groups of samples were fabricated: one with a graded cell size, and another with a graded density, which affects the wall thickness of the gyroid and enables a gradual transition from a porous to a homogeneous material. Each group includes two gradation variants, differing mainly in the ratio of the smallest pore size to wall thickness (Fig. 1).

- Var A - graded cell size in a defined shape 900×525 $\mu$m $\rightarrow$ 600×350 $\mu$m $\rightarrow$ 900×525 $\mu$m
- Var B - graded cell size in a defined shape 600×350 $\mu$m $\rightarrow$ 900×525 $\mu$m $\rightarrow$ 600×350 $\mu$m

- Var C - graded density in a defined shape 80 % density → 40 % density (cell 900 × 525 $\mu$m) → 80 % density
- Var D - graded density in a defined shape 80 % density → 40 % density (cell 600 × 350 $\mu$m) → 80 % density

The fabrication of the test specimens was carried out in cooperation with the company ProSpon Ltd. (Kladno, Czech Republic). The specimens were produced from Ti6Al4V material using 3D printing technology—Selective Laser Melting (SLM). The printed parts underwent post-processing thermal treatment using the Hot Isostatic Pressing (HIP) method with parameters of 150 MPa pressure, 900°C temperature, and a dwell time of 2 hours.

The thermally stabilized specimens were subsequently divided into two groups. One group was tested in the as-HIPed condition. The second group underwent surface etching to remove incompletely fused powder particles from the surface of the samples [9]. Final mechanical testing was therefore conducted on both groups of specimens—those with a graded cell structure (Fig.2) and those with graded density(Fig. 3)—each including etched and non-etched surfaces.

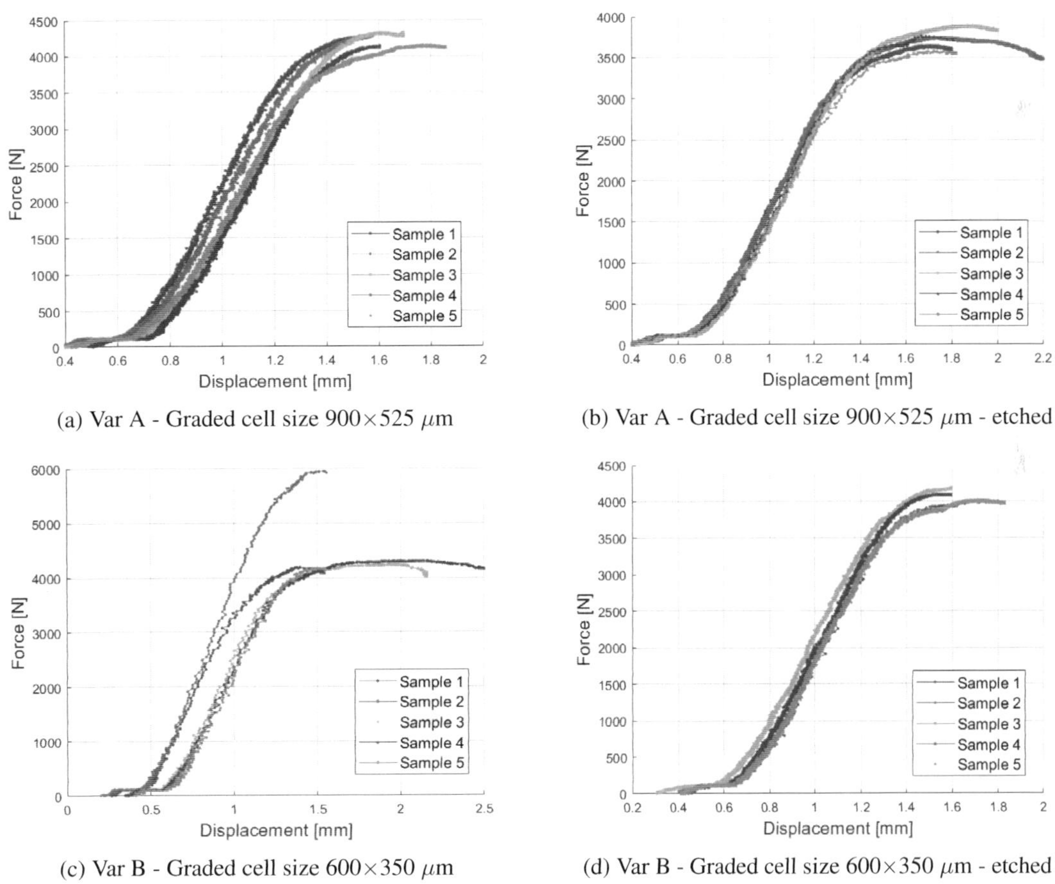


(a) Var A - Graded cell size 900×525 $\mu$m

(b) Var A - Graded cell size 900×525 $\mu$m - etched

(c) Var B - Graded cell size 600×350 $\mu$m

(d) Var B - Graded cell size 600×350 $\mu$m - etched

Fig. 2: Results of uniaxial pressure test on samples with graded cell size.

Mechanical testing was performed using a pneumatic press (LiTeM, Italy) equipped with a cylinder and a force sensor with a maximum load capacity of 7 kN. The specimens were loaded along their longitudinal axis in displacement-controlled mode. The loading rate was 0.05 mm/s, based on the standard ČSN EN ISO 13314:2011 "Mechanical testing of metals – Ductility testing – Compression test for porous and cellular metals." The measured data were recorded at a frequency of 10 Hz into a text file, which was subsequently processed using Matlab software.

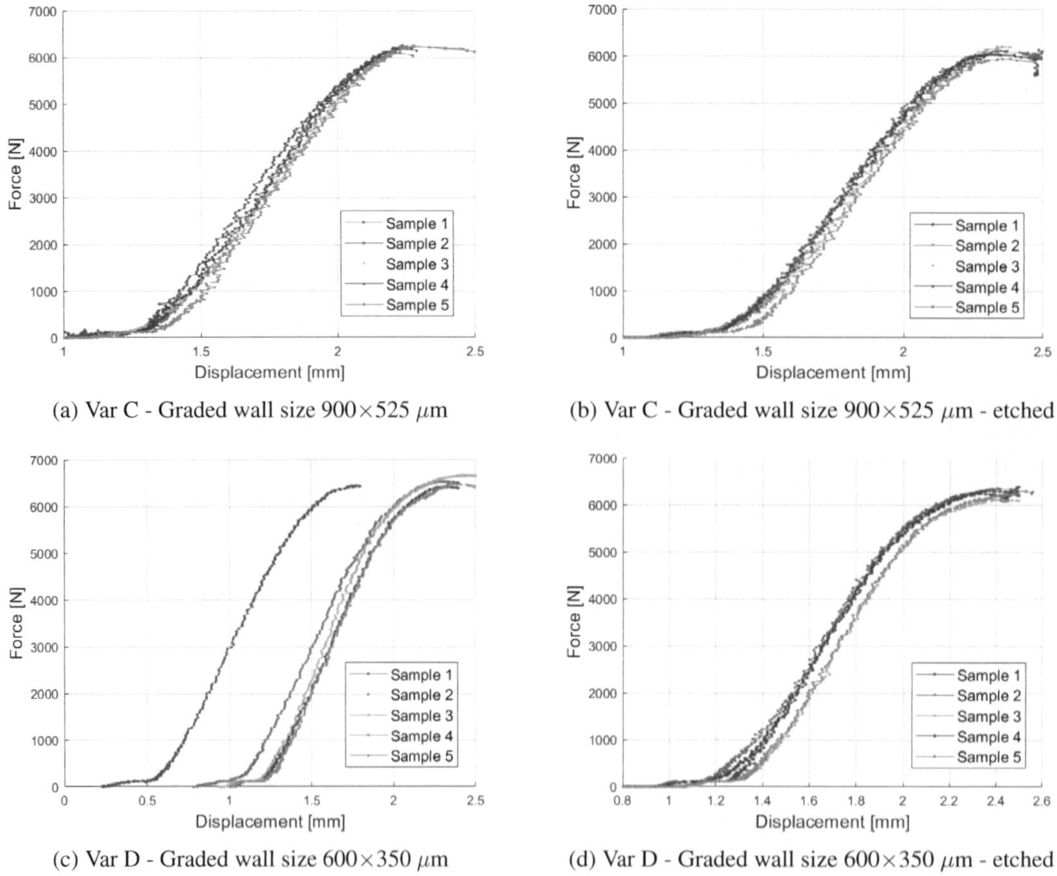


(a) Var C - Graded wall size 900×525 $\mu$m

(b) Var C - Graded wall size 900×525 $\mu$m - etched

(c) Var D - Graded wall size 600×350 $\mu$m

(d) Var D - Graded wall size 600×350 $\mu$m - etched

Fig. 3: Results of uniaxial pressure test on samples with density-induced variation in gyroid wall thickness.

## 3 Results

Experimental data were evaluated using the least squares method. Linear regression was applied to the section of the stress–strain curve that exhibited linear behavior. The yield point was determined as the stress corresponding to 0,2 % plastic compressive strain, using a quasi-elastic gradient defined by the slope of the initial linear portion of the compressive stress–strain curve. Specimens with a graded cell structure withstood uniaxial compressive loads ranging from 3128 to 4125 N, with outliers reaching up to 5931 N prior to significant deformation. In contrast, specimens with a graded wall structure endured compressive loads between 4891 and 6565 N before severe deformation occurred. The highest global elastic modulus was observed in specimens with a graded wall and a cell size of 900×525 $\mu$m,, ranging from 3240 MPa to 3750 MPa. Conversely, the lowest values were recorded in samples with a graded cell size of 600×350 $\mu$m, with modulus between 1800 MPa and 2147 MPa. The results for the global elastic modulus are presented in Fig. 4.

## 4 Conclusion

The results were significantly influenced by initial manufacturing imperfections inherent to the additive manufacturing process. One of the primary limitations is the intrinsic resolution of the process, with typical layer thicknesses ranging from 30 to 40$\mu$m [14]. Additionally, the high melting temperature of titanium alloys promotes partial sintering of the metallic powder, leading to neck formation between particles. While this enhances structural homogeneity, it also introduces minor geometric deviations.

Another contributing factor is the mismatch between the designed unit cell dimensions and the effective volume occupied by the sintered material. Due to the particle size distribution and localized melting phenomena, the actual geometry may diverge from the nominal design, resulting in surface irregularities, porosity gradients,

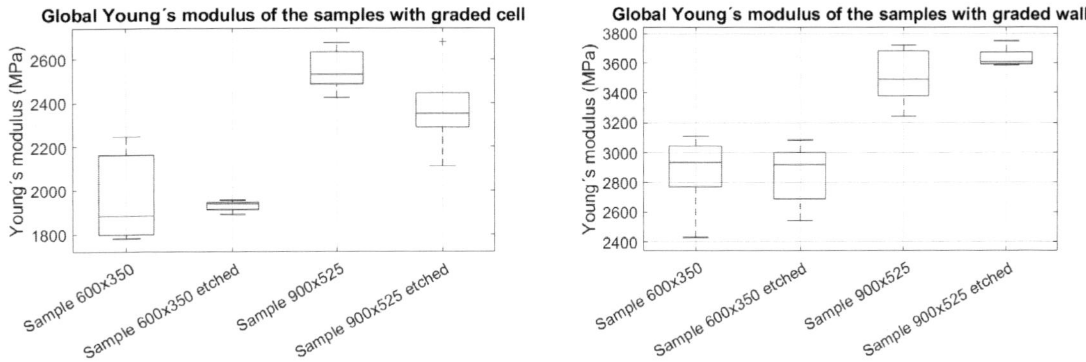


(a) Graded size of the main cell of the gyroid

(b) Graded thickness of the gyroid´s wall

Fig. 4: Trend of the global elastic modulus of the gyroid structure for different porosities and types of grading.

and local inconsistencies.

The predominant failure mechanisms observed included local wall buckling, fracture perpendicular to the loading direction, and shear failure along diagonal planes oriented approximately 45°, as shown in Fig.5 and Fig.6. These observations are consistent with the regions of maximum shear stress. Samples with a graded wall size of $600 \times 350 \mu m$, in both etched and unetched conditions, exhibited a barreling-type deformation. The presence of gradient porosity played a key role in preserving residual structural integrity by enabling stress redistribution even after partial damage, thereby delaying catastrophic collapse.

Fig. 5: Graded cell size $900 \times 525$ $\mu m$ etched.

Fig. 6: Graded wall size $900 \times 525$ $\mu m$ etched.

The optimization of the porosity gradient was identified as a critical factor influencing the mechanical performance of the structures. A well-designed porosity distribution facilitates improved load transfer, increased stiffness, and a more uniform stress distribution, thereby delaying the onset of failure. Additionally, an appropriately engineered gradient contributes to enhanced overall strength and improved residual structural stability following partial damage.

Future research will focus on quantitatively correlating the degree of geometric deviation with mechanical performance across various porosity gradients. Advanced imaging techniques such as micro-computed tomography ($\mu$CT) will be employed to capture internal defects and microstructural variations with higher fidelity.

## Acknowledgement

We are thankful for financial support provided by The Czech Health Reserch Council (AZV) No. NW24-1000118 and Czech Technical University in Prague No. SGS23/152/OHK1/3T/11.

## References


[1] G. Baker, Osseointegrated implants for dental rehabilitation following ablative and reconstructive surgery, Operative Techniques in Otolaryngology-Head and Neck Surgery, 4 (1993) 115-122, doi: 10.1016/S1043-1810(10)80137-X.

[2] M. Jaber, et al. Osseointegrated implants for dental rehabilitation following ablative and reconstructive surgery, Frontiers in Bioengineering and Biotechnology, 9 (2022) doi: 10.3389/fbioe.2022.995266.

[3] A. Bandyopadhyay, et al. Improving biocompatibility for next generation of metallic implants, Progress in Materials Science, 101053 (2023) doi: 10.1016/j.pmatsci.2022.101053.

[4] L. E. Murr, et al. Metal Fabrication by Additive Manufacturing Using Laser and Electron Beam Melting Technologies, Journal of Materials Science & Technology, (2012) 1–14 doi: 10.1016/S1005-0302(12)60016-4.

[5] A. Jíra, et al. Mechanical Properties of Porous Structures for Dental Implants: Experimental Study and Computational Homogenization, Materials 14 (2021) 1–25, doi: 10.3390/ma14164592.

[6] Z. Faqihah, et al. Biomechanical Effects of the Porous Structure of Gyroid and Voronoi Hip Implants: A Finite Element Analysis Using an Experimentally Validated Model, Materials, 4 (2023) 3298 doi: 10.3390/ma16093298.

[7] L. C. Zhang, H. Attar. Selective Laser Melting of Titanium Alloys and Titanium Matrix Composites for Biomedical Applications: A Review, Advanced Engineering Materials, 18 (2016) 463–475, doi: 10.1002/adem.201500419

[8] E. Yang, et al. Effect of geometry on the mechanical properties of Ti-6Al-4V gyroid structures fabricated via SLM: A numerical study, Materials and Design, (2019) 1–24, doi: 10.1016/j.matdes.2019.108165.

[9] T. S. Tshephe, et al. Additive manufacturing of titanium-based alloys- A review of methods, properties, challenges, and prospects, Heliyon, 8 (2022) 9–41, doi: 10.1016/j.heliyon.2022.e09041.

[10] P. Vakrčka, et al. Mechanical testing and numerical modelling of porous structures improving oseintegration of implants, Acta Polytechnica CTU Proceedings, 26 (2020) 127-133, doi: 10.14311/APP.2020.26.0126.

[11] F. Danielli et al. PCL strut-like scaffolds appear superior to gyroid in terms of bone regeneration within a long bone large defect: An in silico study, Journal of the Mechanical Behavior of Biomedical Materials 1, (2025) 107022, doi: 10.1016/j.jmbbm.2025.107022.

[12] M. Frolo, et al. Biomechanical Analysis of Palateless Splinted and Unsplinted Maxillary Implant-Supported Overdentures: A Three-Dimensional Finite Element Analysis, Materials, 16 (2023) 1-12, doi: 10.3390/ma16155248.

[13] M. Niinomi, Mechanical properties of biomedical titanium alloys, Materials Science and Engineering, 234 (1998) 231-–236, doi: 10.4236/ampc.2015.57025.

[14] K.M. Rajan et al., A review on various approaches of 3D printing of Ti-Alloy, Materials Today: Proceedings, (2022) 3865-3868, doi: 10.1016/j.matpr.2022.04.532.

# Influence of Etching on the Mechanical Properties of Thin 3D Printed Ti6Al4V Alloy Specimens

P. Bočan[1,*], A. Jíra[1]

[1] *Czech Technical University in Prague, Faculty of Civil Engineering, Department of Mechanics, Thákurova 7,166 29 Prague 6, Czech Republic*

[*] *petr.bocan@fsv.cvut.cz*

**Abstract:** This study investigates the effect of surface etching on the mechanical performance of thin 3D-printed Ti6Al4V specimens subjected to three-point bending. Three geometries, all post-processed using Hot Isostatic Pressing (HIP), were tested, with half additionally undergoing chemical etching. Flexural strength and apparent fracture toughness were evaluated. The results indicate that surface etching effectively removes partially melted powder particles from the sample surface but does not significantly influence the mechanical properties at the tested thicknesses. These findings contribute to the optimization of surface treatment strategies for load-bearing biomedical implants.

**Keywords:** Ti6Al4V alloy; Three-point bending experiment; modified SENT geometry; Flexural strength; Fracture toughness

## 1 Introduction

3D printing technology is one of the most advanced additive manufacturing methods used in various industries. It enables the fabrication of complex structures and drives innovation, making it a highly researched and adopted manufacturing technique. In the biomedical field, 3D printing plays a critical role in the development of intraosseous implants. This technology allows for the creation of porous surfaces on implants, which promote osseointegration by enabling bone cells to grow into the pores, thereby strengthening the interface between the bone and the implant.

The primary challenges associated with 3D printed titanium powder alloys involve manufacturing inaccuracies, such as structural imperfections, partially melted powder balls on the surface, and weak interconnections between structural elements [1]. These issues are particularly critical in thin components in porous structures. Several studies [2, 3, 4] have examined these imperfections in 3D printed Ti6Al4V alloy samples, with a focus on thin porous geometries, highlighting the geometric complexities and limitations that arise during the additive manufacturing process.

3D printing often requires extensive post-processing of fabricated samples. Due to the nature of the manufacturing process, micropores can form between partially melted powder particles, leading to morphological defects and reduced mechanical strength. Additionally, the high temperatures involved during fabrication induce residual stresses, which may result in the formation of microcracks. These issues can be mitigated through appropriate post-processing techniques, with annealing and Hot Isostatic Pressing (HIP) being the two primary methods. Another common defect is the presence of insufficiently melted powder particles adhered to the sample surface, which can be effectively removed by chemical etching.

The purpose of the present study was to evaluate the mechanical properties of thin Ti-6Al-4V samples under bending loads. Three different specimen geometries, all based on modified single edge notched tension (SENT)-type designs, were tested to assess both flexural strength and apparent fracture toughness using a three-point bending setup. To eliminate internal stresses and consolidate internal pores, all samples underwent post-processing via Hot Isostatic Pressing (HIP) [5]. In addition, half of the samples were subjected to surface etching to remove partially melted powder particles that remained on the surface.

## 2 Specimen design and experimental procedure

Ti6Al4V alloy samples were fabricated using Selective Laser Sintering (SLS) and subsequently post-processed via Hot Isostatic Pressing (HIP). The HIP process involved compressing the sample in a high-temperature chamber at 150 MPa and 900 °C for 2 hours. The specimens were designed as beams with a central notch, based on modified single notched tension (SENT) samples. Three cross-sectional groups were prepared: 1 × 4 mm, 1.5 × 6 mm, and 3 × 3 mm, all with a constant length of 50 mm. The notch dimensions were scaled to maintain a constant height-to-width ratio of 0.15. The geometry and dimensions of all tested specimens are shown in Figure 1.

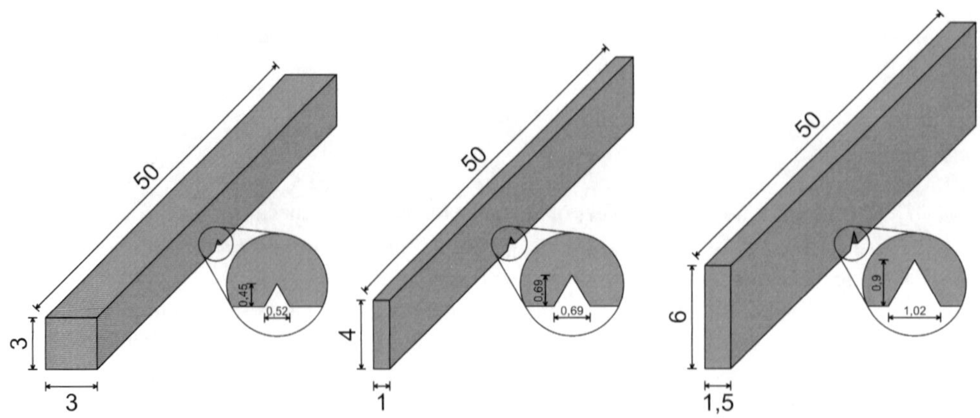


Fig. 1: Examples of individual geometries of the modified SENT-type samples with corresponding basic dimensions.

A common issue with 3D printed samples is the presence od partially unmelted powder balls adhered to the surface after fabrication. One effective method for removing these surface particles is chemical etching, in which the samples are immersed in a chemical solution that dissolves and removes excess material. Figure 2 presents two SEM images: the left image shows a sample surface before chemical etching, while the right image displays the surface after treatment, illustrating the effectiveness of the etching process.

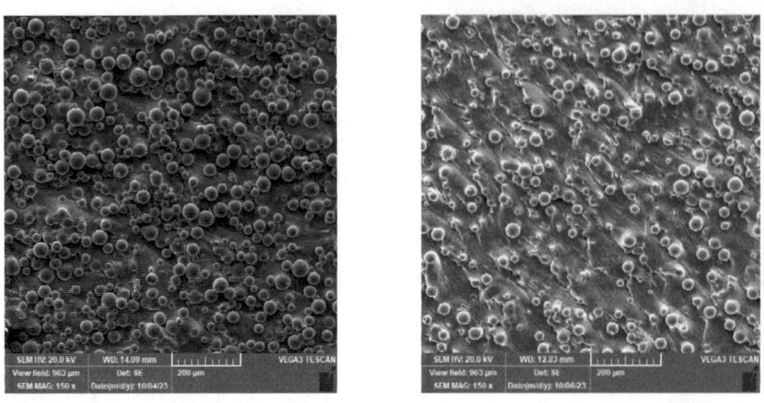

Fig. 2: Illustration of the effect of chemical etching to remove powder beads from the sample surface, with the left image showing the surface before etching and the right image after etching.

## 3 Results and discussion

A three-point bending test was conducted at a loading rate of 10 mm/s, using a support span of 40 mm. The applied load was measured using a MARK-10 press, supplemented by a macro-scale 3D Digital Image Correlation (DIC) system. The experimental data, including beam deflection under load, were post-processed to generate load-distance diagrams, as shown in Figure 3.

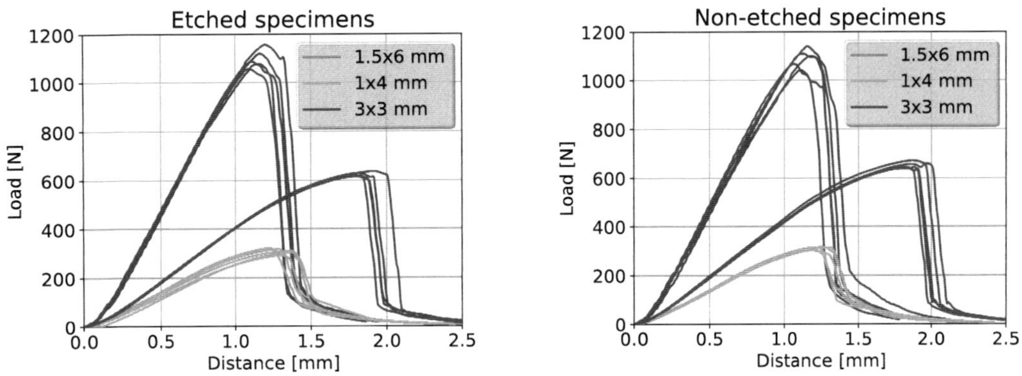


Fig. 3: Load-distance diagram showing the dependence of force on displacement for each specimen.

Prior to testing, the dimensions of all specimens were measured, with particular attention given to the width ($W$) and thickness ($B$). The maximum applied force ($F_{max}$) for each specimen was then determined from the load-displacement diagrams. These variables are summarized in Table 1.

The ultimate flexural strength ($\sigma_f$) was calculated using the following equation:

$$\sigma_f = \frac{3F_{max}L}{2B(W-a)^2} \tag{1}$$

where $L$ is the support span (constant at 40 mm) and $a$ is the crack length, which was kept constant for all geometries, as indicated in Table 1.

The apparent fracture toughness ($K_Q$) was determined according to ASTM E399-22 [6], using the following expression:

$$K_Q = \frac{F_{max}L}{BW^{\frac{3}{2}}} \cdot f_{\left(\frac{a}{W}\right)} \tag{2}$$

where $f\left(\frac{a}{W}\right)$ is a dimensionless geometry function defined as:

$$f\left(\frac{a}{W}\right) = 3\left(\frac{a}{W}\right)^{1/2} \cdot \frac{1.99 - \left(\frac{a}{W}\right)\left(1-\frac{a}{W}\right)\left[2.15 - 3.93\left(\frac{a}{W}\right) + 2.7\left(\frac{a}{W}\right)^2\right]}{2\left(1+2\frac{a}{W}\right)\left(1-\frac{a}{W}\right)^{3/2}} \tag{3}$$

A summary of the measured dimensions and the corresponding calculated results is provided in Table 1, while a graphical representation of the flexural strength and apparent fracture toughness values is shown in Figure 4.

Tab. 1: Summary of basic values for calculation of mechanical parameters for etched and unetched samples.

| | $a$ [mm] | $B$ [mm] | $W$ [mm] | $F_{max}$ [N] | $\sigma_f$ [MPa] | $K_Q$ [MPa$\sqrt{\text{m}}$] |
|---|---|---|---|---|---|---|
| **Etched specimens** | | | | | | |
| 1.5x6 mm | 0.9 | 1.50 ± 0.01 | 6.11 ± 0.03 | 1115.0 ± 36.88 | 1636.20 ± 54.12 | 62.53 ± 2.3 |
| 1x4 mm | 0.6 | 1.00 ± 0.01 | 4.01 ± 0.06 | 317.0 ± 10.14 | 1633.00 ± 52.23 | 50.62 ± 1.2 |
| 3x3 mm | 0.45 | 3.00 ± 0.02 | 3.05 ± 0.03 | 632.2 ± 7.22 | 1867.92 ± 21.34 | 50.45 ± 1.2 |
| **Non-etched specimens** | | | | | | |
| 1.5x6 mm | 0.9 | 1.53 ± 0.02 | 6.17 ± 0.07 | 1109.20 ± 35.18 | 1568.25 ± 49.74 | 60.14 ± 1.7 |
| 1x4 mm | 0.6 | 1.03 ± 0.01 | 3.99 ± 0.01 | 319.20 ± 5.84 | 1515.14 ± 27.74 | 50.28 ± 1.2 |
| 3x3 mm | 0.45 | 3.03 ± 0.02 | 3.10 ± 0.02 | 658.60 ± 11.52 | 1860.28 ± 32.53 | 50.47 ± 0.9 |

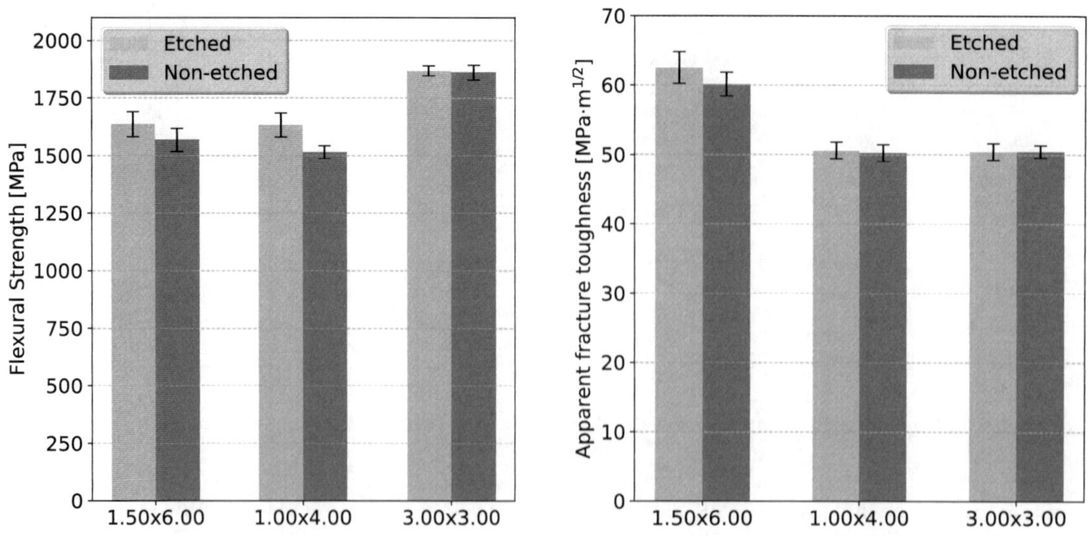


Fig. 4: The summary of the mechanical properties retrieved by experimental data.

# 4 Conclusion

This research focused on the mechanical testing of thin Ti6Al4V specimens under three-point bending test. All specimens were post-processed using Hot Isostatic Pressing (HIP) to reduce internal porosity, improve material integrity and reduce initial stresses. To assess the influence of surface condition on mechanical behaviour, both chemically etched and non-etched specimens were tested. This comparison aimed to reveal potential differences in flexural strength $\sigma_f$ and apparent fracture toughness $K_Q$ resulting from the removal of partially melted surface particles and the improvement of surface quality.

The specimen geometries were based on SENT-type configurations with slight modifications to accommodate the experimental setup. Three different cross-sectional dimensions were tested. Due to the reduced thickness of the specimens, the conditions necessary to achieve a plane strain state - required to determine the true stiffness of the fracture $K_{IC}$ - were not satisfied. Consequently, the specimens likely exhibited a plane stress state, and the calculated fracture toughness values are referred to as apparent fracture toughness $K_Q$.

The research demonstrated that surface treatment by chemical etching did not have a significant influence on the mechanical properties of the specimens. The evaluated values showed only minor differences, all within the range of standard deviations. Therefore, it can be concluded that chemical etching does not significantly affect the mechanical performance for the tested thicknesses. However, issues may arise with thinner specimens, where etching can substantially reduce the cross-sectional area, potentially leading to a decrease in load-bearing capacity.

# Acknowledgment


The financial support from the Technology Agency of the Czech Republic (project No. TQ15000166) and the Czech Technical University in Prague (SGS project No. SGS23/152/OHK1/3T/11) is gratefully acknowledged.


# References


[1] Murchio, S., et al. Additively manufactured Ti–6Al–4V thin struts via laser powder bed fusion: Effect of building orientation on geometrical accuracy and mechanical properties. Journal of the Mechanical Behavior of Biomedical Materials 119 (2021) 104495, doi: 10.1016/j.jmbbm.2021.104495.

[2] A. Jíra, et al. Mechanical Properties of Porous Structures for Dental Implants: Experimental Study and Computational Homogenization, Materials 14 (2021) 1–25, doi: 10.3390/ma14164592.

[3] L. Řehounek, et al. Optimization of 3D printed trabecular structures for implantology and their mechanical analysis, (2017). Unpublished report from TAČR project No. TJ01000328.

[4] Sereshk, M. R. V., Faierson, E., Study of defects and dimensional accuracy to determine manufacturability thresholds for laser powder bed fusion of Ti-6Al-4V lattices. The International Journal of Advanced Manufacturing Technology 123 (2022) 3795-3806, doi: 10.1007/s00170-022-10325-w.

[5] YAN, Xingchen, et al. Microstructural and mechanical optimization of selective laser melted Ti6Al4V lattices: Effect of hot isostatic pressing. Journal of Manufacturing Processes, 2022, 77: 151-162, doi: 10.1016/j.jmapro.2022.02.024

[6] ASTM E399-22: Standard Test Method for Linear-Elastic Plane-Strain Fracture Toughness of Metallic Materials (2022) West Conshohocken, PA, ASTM International.

# Compression Behavior of 3D-Printed stochastic Structures at Varying Densities

A. Boualleg[1, a], D. Cirkl [1]

[1] *Technical University of Liberec; Czech Republic*
[a] *abdelmadjid.boualleg@tul.cz*

**Abstract:** Voronoi-based cellular structures have attracted growing interest due to their ability to combine low weight with complex geometry and tunable mechanical performance. In this study, we investigate the compressive behavior of 3D-printed Voronoi structures fabricated with different densities. Using stereolithography (SLA) 3D printing and a viscoelastic photopolymer, a set of samples with varying internal configurations was produced and subjected to quasi-static uniaxial compression tests. The energy absorption was evaluated with respect to density. The results show a clear correlation between increased relative density and improved load-bearing capacity and energy absorption. Despite their randomness, Voronoi structures exhibit consistent deformation modes influenced by local cell morphology. Our findings highlight the potential of Voronoi lattices for load bearing and energy absorption.

**Keywords:** 3D printing; 3D Voronoi; flexible material; Energy.

## 1 Introduction

Architected cellular materials have attracted increasing attention in recent years due to their ability to provide optimized combinations of stiffness, strength, energy absorption, and lightweight performance. Among these, Voronoi-based lattice structures represent a class of stochastic geometries inspired by natural cellular forms such as bone, wood, and coral. Unlike periodic lattices, Voronoi structures offer irregular topologies that can reduce stress concentrations and enable better mechanical performance under complex loading conditions [1]. Their design flexibility allows the control of local density and deformation modes, which is especially useful in impact mitigation and energy-absorbing applications. With the development of advanced Additive Manufacturing (AM) techniques, particularly stereolithography (SLA), it is now possible to fabricate these complex structures with high precision using photopolymeric materials that exhibit viscoelastic behavior [2, 3]. Several studies have shown that the mechanical performance of such lattices can be tuned by adjusting design parameters like strut diameter, cell size, and relative density [3]. However, understanding the influence of these parameters, especially under compressive loading, remains an active area of research. This study aims to investigate the compressive behavior and energy absorption capacity of SLA-fabricated Voronoi structures with different relative densities.

## 2 Methodology

### 2.1 Generate the structure

The Voronoi algorithm was used to generate 3D lattice structures by partitioning a predefined volume based on the proximity of randomly distributed seed points. Each seed defines a cell, and the interfaces between cells form the struts of the lattice. The structures examined in this article was created with 80 cells and variable strut diameters (0.7 mm, 1 mm, and 1.3 mm). Initial tessellation was performed in MATLAB, and short struts (<2 mm), which negatively affect mechanical behavior, were removed by merging their endpoints at the midpoint. The resulting geometry was then converted into solid 3D structures, with strut data exported via Python for analysis. The discontinuities at vertices were resolved using the "Unite Solids" tool in MARC Mentat Software, and connecting spheres was added at junctions to ensure smooth transitions between struts. The finalized structure was exported as an STL file and fabricated using vat photopolymerization, enabling experimental validation of the computational model.

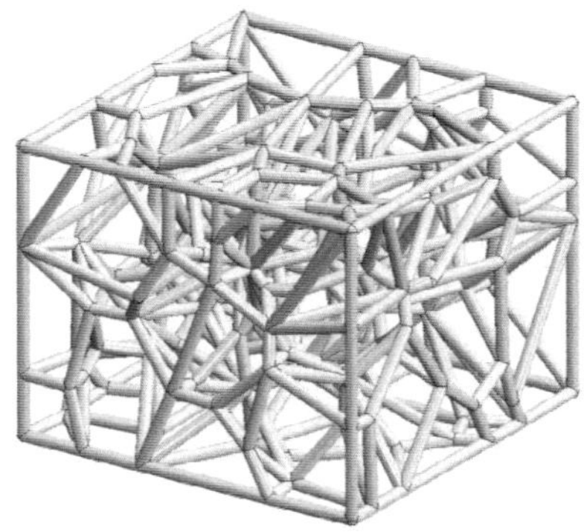

Fig. 1: 3D Voronoi structure generated model.

## 2.2 Fabrication of 3D printing samples and experiments

### 2.2.1 Printing samples

The dog-bone samples was fabricated via vat photopolymerization to investigate the effect of print orientation on mechanical properties. Samples was printed in three orientations (X, Y, and Z) as shown in Figure 2 using masked stereolithography (MSLA) on a Prusa SL1S printer. Key printing parameters included a 0.05 mm layer height, a 100-second initial exposure, and a 20-second layer exposure. After printing, parts was cleaned in isopropyl alcohol (IPA) to remove uncured resin and post-cured for 15 minutes to ensure complete polymerization. The same printing setup was used to fabricate three Voronoi structures, each consisting of 80 cells but differing in strut diameter: 0.7 mm, 1 mm, and 1.3 mm, allowing for evaluation of structural behavior under consistent manufacturing conditions.

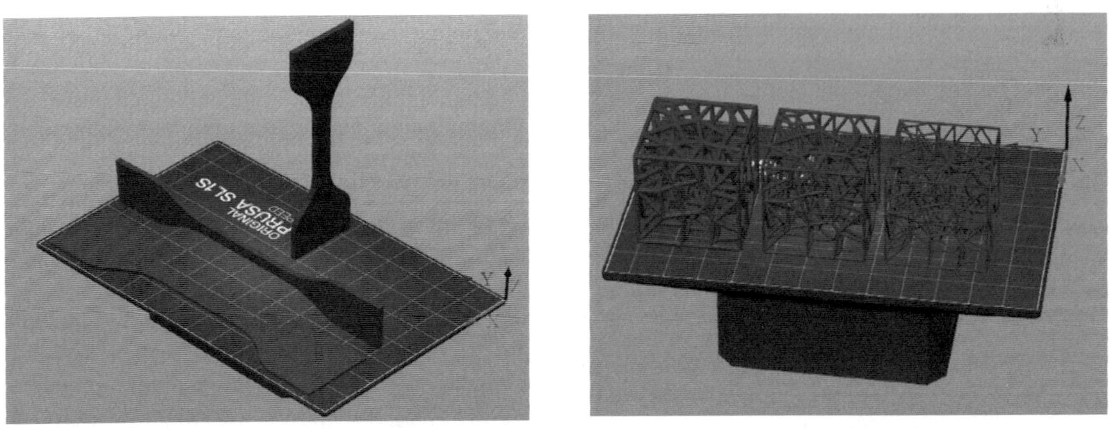


a) The three tensile test samples are presented in three directions, X, Y, and Z.

b) The three 3D Voronoi structures with three different strut diameters.

Fig. 2: 3D printing Samples of tensile test and Voronoi structure.

### 2.2.2 Experimental Procedure for Tensile and Compression Tests

Tensile tests were conducted on samples designed according to the standard ASTM D412 [4], see Figure 2(a). To evaluate the influence of printing orientation on the mechanical properties, samples was printed in three different directions (X, Y, and Z) and tested accordingly. For each orientation, three virgin samples were tested, and the average results was reported.

Compression tests was performed to investigate the mechanical performance of 3D-printed Voronoi structures, see Figure. 2(b). Each virgin specimen was subjected to a single cycle of loading and unloading up to a maximum strain of 0.6. The tests was carried out on structures with three different unit cell densities: 0.7 mm, 1 mm, and 1.3 mm. A constant deformation rate of 50 mm/min was applied. Each configuration was tested three times, and the average values were calculated.

# 3 Results

Tensile tests conducted on samples printed in the X, Y, and Z directions using the SLS process with Ultrasint® BASF EL 60 material showed minimal variation in mechanical performance as presented in the Figure 3. The force-displacement responses across orientations were highly consistent, indicating that the printing direction had a negligible effect. This isotropic behavior is largely attributed to the characteristics of the MSLA process, where each layer is cured simultaneously using a uniform light mask, ensuring consistent polymerization across the build plane. Additionally, Ultrasint® BASF EL 60 is formulated to offer high elasticity and uniform cross-linking, contributing to reliable interlayer adhesion and mechanical homogeneity throughout the printed part [6]. These properties help minimize anisotropy, making the mechanical response of printed parts largely independent of orientation.

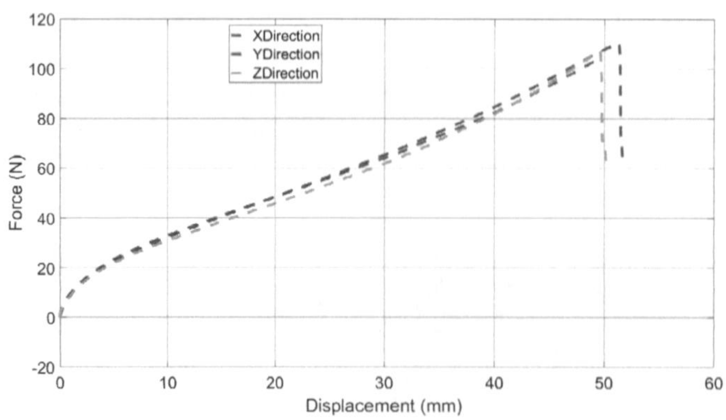


Fig. 3: Tensile test results of three directions X, Y, and Z.

The compression behavior of 3D Voronoi lattice structures with varying strut diameters (0.7 mm, 1.0 mm, and 1.3 mm) was experimentally investigated. The force–displacement curves and the corresponding absorbed energy responses are presented in Figure 4. From the force–displacement plots, it is evident that the mechanical response of the structures is significantly influenced by strut diameter. All structures exhibit an initial buckling elastic region followed by a plateau, which is characteristic of cellular materials undergoing progressive collapse. Beyond the plateau, a densification phase occurs where the force increases sharply.

The structure with the smallest strut diameter (0.7 mm), Figure 4 (a), showed the lowest peak and plateau forces, indicating reduced load-bearing capacity. This behavior is expected, as the reduced cross-sectional area directly decreases the load-bearing capacity of the structure. The 1.0 mm strut diameter structure, Figure 4 (b), showed an intermediate behavior, while the structure with the largest strut diameter (1.3 mm), Figure 4 (c), demonstrated a substantial increase in both peak force and overall stiffness. The densification onset for all samples occurred at approximately the same displacement (12 mm), suggesting that the global deformation characteristics remain similar despite changes in strut diameter. However, the magnitude of the force in this region varies significantly, again highlighting the role of strut thickness in structural rigidity.

The absorbed energy during the loading phase, shown in Figure 4(d), was calculated using the energy absorption formula as presented in Eq. (1):

$$EA = \int_{\delta_{min}}^{\delta_{max}} F(\delta)\, d\delta \qquad (1)$$

Where $EA$ is the absorbed energy in (mJ), $F(\delta)$ is the applied load as a function of displacement. $\delta_{max}$, $\delta_{min}$ are the maximum and minimum of the displacement, respectively. This integral represents the area under the load displacement curve during the loading phase, corresponding to the energy absorbed by the structure before unloading.

As shown in Figure 4(d), the absorbed energy increases nonlinearly with displacement for all strut diameters. The structure with a 1.3 mm strut diameter absorbed the highest energy, reaching over 600 mJ at the maximum displacement. The specimen with a 1.0 mm strut diameter exhibited intermediate energy absorption, while the 0.7 mm structure absorbed less than 100 mJ. These results confirm that increasing strut diameter enhances the energy absorption capability of the structure, indicating that thicker struts provide greater stiffness and resistance to deformation. Consequently, structures with larger strut diameters are more suitable for applications requiring improved impact resistance.

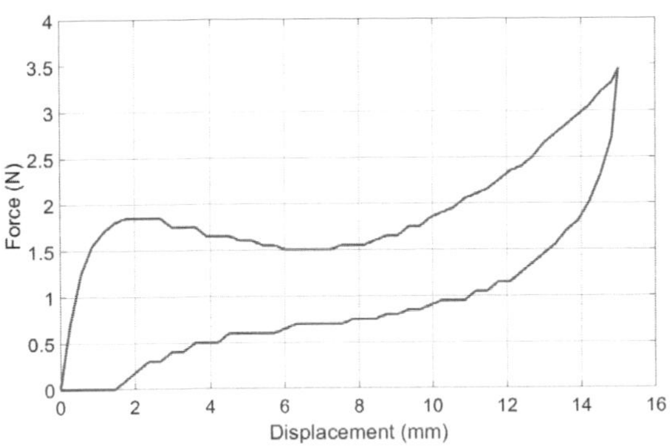


a) Strut diameter of 0.7 mm.

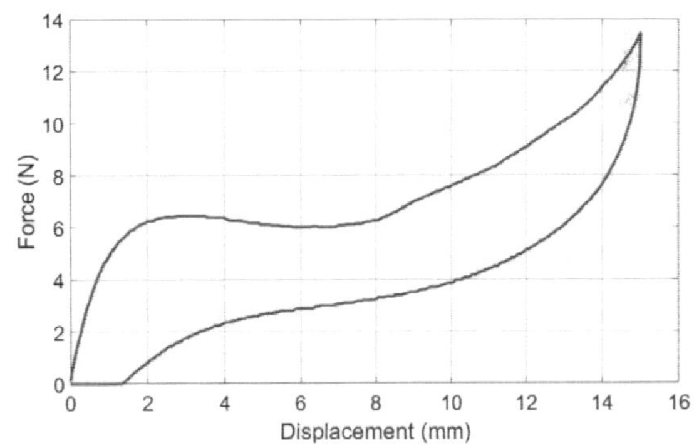


b) Strut diameter of 1 mm.

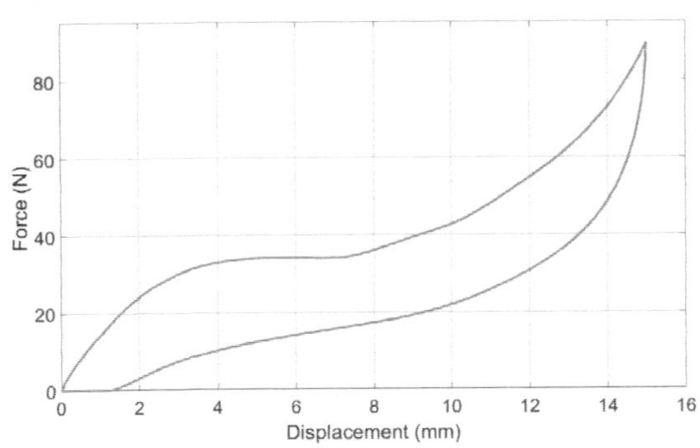


c) Strut diameter of 1.3 mm.

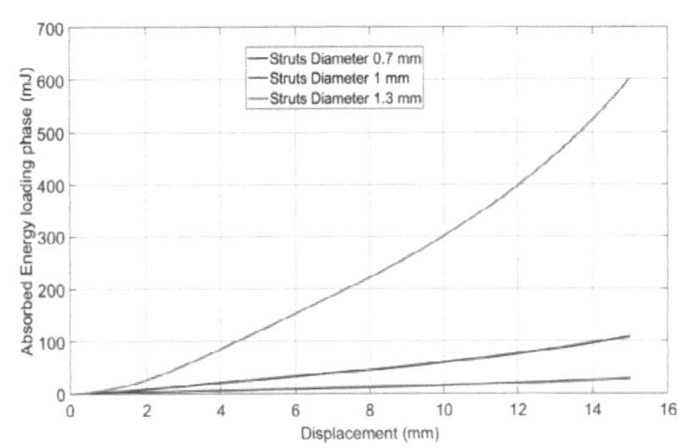


d) Absorbed energy of loading phase.

Fig. 4: Force-Displacement curves of compression test at 50 mm/min with various 3D Voronoi strut diameters.

These results demonstrate a strong positive correlation between strut diameter and both load-bearing and energy absorption capacity. The findings highlight a trade-off in structural design: while thinner struts may offer weight savings and increased porosity for applications such as tissue engineering scaffolds, thicker struts provide superior mechanical strength and energy absorption, which are beneficial for structural and protective applications.

## 4 Conclusion

Tensile testing across X, Y, and Z orientations showed negligible differences in mechanical response, indicating near-isotropic behavior. This consistency is attributed to the characteristics of the MSLA printing process and the material's formulation, which promotes uniform cross-linking and interlayer adhesion.

This study also explored the compressive performance of 3D-printed Voronoi cellular structures with varying strut diameters, highlighting how geometric tuning influences mechanical behavior. The experimental results demonstrated that increasing the strut diameter significantly enhances both the load-bearing capacity and energy absorption, despite all samples exhibiting similar overall deformation patterns. Structures with thicker struts (1.3 mm) showed higher stiffness, greater peak force, and substantially improved energy absorption compared to those with thinner struts 1 and 0.7 mm.

The observed trends confirm the potential of Voronoi-based designs for applications where mechanical strength and energy absorption are critical, such as protective components or load-bearing lightweight structures. At the same time, the ability to control porosity and density through strut thickness makes these architectures attractive for biomedical applications, where a balance between mechanical integrity and permeability is essential.

## Acknowledgement

## References

[1] I. Maskery et al., Compressive failure modes and energy absorption in additively manufactured double gyroid lattices, Addit. Manuf., vol. 16, pp. 24–29, 2017, doi: 10.1016/j.addma.2017.04.003.

[2] T. D. Ngo et al., A review of materials, methods, applications and challenges," Compos. Part B Eng., vol. 143, no. December 2017, pp. 172–196, 2018, doi: 10.1016/j.compositesb.2018.02.012.

[3] Y. Wu et al., Piezoelectric materials for flexible and wearable electronics: A review, Mater. Des., vol. 211, p. 110164, 2021, doi: 10.1016/j.matdes.2021.110164.

[4] Gümrük R, Mines RAW, Karadeniz S., Static mechanical behaviours of stainless steel micro-lattice structures under different loading conditions," Mater. Sci. Eng. A, vol. 586, pp. 392–406, 2013, doi: 10.1016/j.msea.2013.07.070.

[5] C. B. Manufacturing, C. R. Specimens, and T. Determina-, "Standard Test Methods for Vulcanized Rubber and Thermoplastic Elastomers —," vol. 16, no. Reapproved, pp. 1–14, 2009, doi: 10.1520/D0412-16R21.2.

[6] Retrieved from: https://forward-am.com/

# Comparison of Incremental and DIC-Based Determination of the Cyclic Stress-Strain Curve for SLM AlSi10Mg

J. Cienciala [1,a] , M. Fusek [1] , R. Halama[1] , M. Mánek[1] , J. Šimota [2] , Z. Pitrmuc [2]

*[1] VSB – Technical University of Ostrava, 17. listopadu 2172/15, 708 00 Ostrava, Czech Republic*
*[2] Czech Technical University in Prague, Technická 4, 160 00 Prague, Czech Republic*
*[a] jakub.cienciala@vsb.cz*

**Abstract:** This study compares three approaches to determining the cyclic stress–strain curve (CSSC) of the additively manufactured AlSi10Mg alloy produced by selective laser melting. The first method follows the incremental step test proposed by Klesnil and Lukáš, while the second is based on digital image correlation (DIC) using the methodology of Halama et al. (2024, DOI: 10.1016/j.prostr.2024.02.045). DIC was applied during a standard low-cycle fatigue test with high strain amplitude, enabling the CSSC to be evaluated without additional specimens or tests. The DIC-based approach provides full-field strain data with minimal experimental effort, offering substantial savings in material, energy, and time. The latter method follows the conventional way of determining CSSC from a set of low cycle fatigue tests. The comparison of results demonstrates good agreement between the methods and highlights DIC as a viable and resource-efficient alternative for CSSC determination in additively manufactured materials.

**Keywords:** Low-cycle fatigue; digital image correlation; incremental fatigue test; cyclic stress-strain curve.

## 1 Introduction

Additive manufacturing using the selective laser melting (SLM) method has become an important tool in recent years for producing geometrically complex metal components. Understanding the behaviour of additively fabricated metals under cyclic loading becomes essential [1]. One of the fundamental tools for analysing the fatigue life of materials is the cyclic stress–strain curve (CSSC), which describes the relationship between stress amplitude and strain amplitude.

This study focusses on comparing two accelerated methods to determine the CSSC of the AlSi10Mg alloy produced by SLM. The first method is based on the incremental test procedure developed by Klesnil and Lukáš [2]. The second method, based on the approach proposed by Halama et al. [3], uses digital image correlation (DIC) applied to tubular specimens during a standard low cycle fatigue (LCF) test with constant strain amplitude. This technique enables simultaneous strain amplitude measurement at various cross sections of the specimen, significantly reducing the total time required to obtain the CSSC. The purpose of this study is to compare the results of both methods applied to the same additively manufactured specimens with conventional CSSC (the third method).

## 2 Material and Production

All specimens were fabricated using the SLM technique with a vertical build orientation from a virgin powder. The samples were printed on a Concept Laser M2 printer using the laser powder bed fusion process with a chessboard scanning strategy, optimized by the manufacturer to ensure uniform heat distribution and minimize distortion. The printing parameters varied for different regions: the outer "skin" was processed with 200 W laser power, 800 mm/s scan speed, 0.112 mm hatch spacing, 140 µm spot size, and 0.025 mm layer height; the "core" used 370 W power, 1400 mm/s speed, 0.112 mm hatch spacing, 190 µm spot size, and 0.050 mm layer height.

After printing, the specimens underwent a stress-relief heat treatment at 300 °C for 2 hours to reduce residual stresses induced during the manufacturing process, followed by furnace cooling. Following heat treatment, the gauge section of each specimen was gently machined to remove surface irregularities while preserving the integrity of the near-surface microstructure [5].

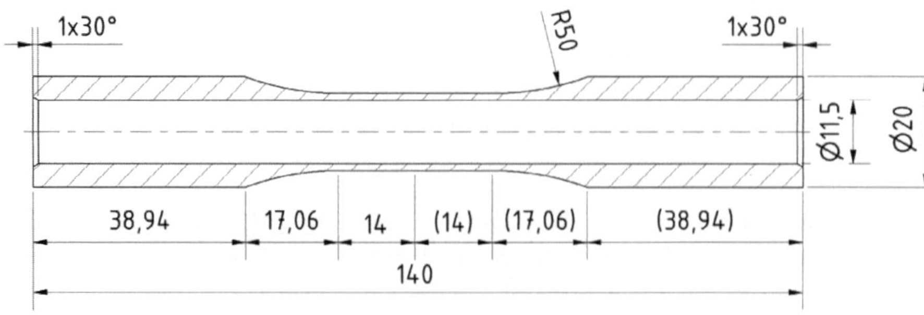


Fig. 1: Scheme of specimen geometry.

## 3 Incremental Testing

All experiments were carried out using the LabControl 100kN/1000Nm hydraulic axial-torsion testing machine. The first approach used to determine the cyclic stress–strain curve was the uniaxial incremental step test [2]. The test was performed under total strain amplitude control, applying a constant strain rate of 0.5%/s. The strain amplitude was gradually increased in discrete steps: 0.25%, 0.35%, 0.5%, 0.75%, and further up to 2% using a fixed increment of 0.25% (Fig. 2a). The material shows slight cyclic softening in the first three levels of strain amplitude, then it seems to be cyclically stable, see Fig. 2b.

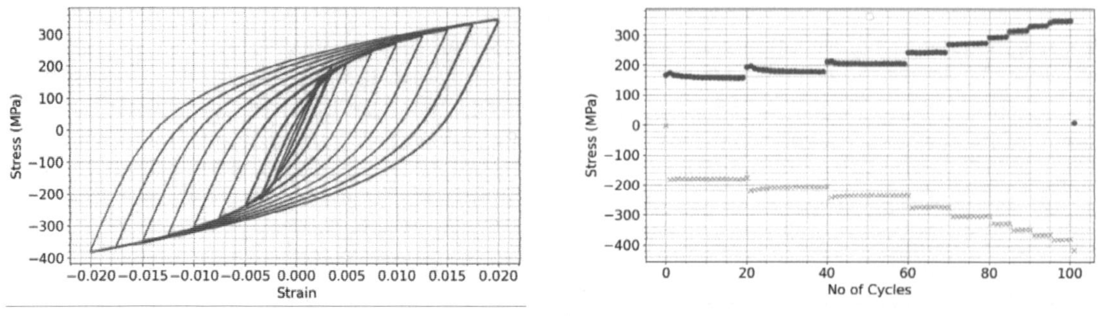


a) Hysteresis loops from the incremental test

b) Maximal and minimal stress evolution

Fig. 2: Chosen results of the uniaxial incremental fatigue test.

## 4 DIC-Based Technique

The accelerated testing technique has been proven on three loading paths considering the same material (using recycled powder) in paper [3]. The mean strain in the gauge portion and the amplitude of the axial strain are maintained constant during the fatigue test in a typical uniaxial strain-controlled fatigue test. Nonetheless, the specimen's cross-sectional area is greater in the rounded portion, which provides the lower stress and strain magnitudes. The digital image correlation approach can be used to optically measure the strain history on the specimen's curved portion. In this study, two CCD cameras with a resolution of 2.3 Mpx and a frame rate of 5 (fps) were used in a 3-D measurement arrangement (Mercury MS software), see Fig. 3a (on left). A visualisation of the DIC strain contours along the specimen is presented in Fig. 3b. Three blue-marked points representing the global coordinate system are visible in the figure. The LCF test was performed at 1.0 % strain amplitude. Harmonic loading was applied with a frequency of 0.125 Hz. Subsequently, the cyclic stress-strain curve was determined using a new custom Python code.

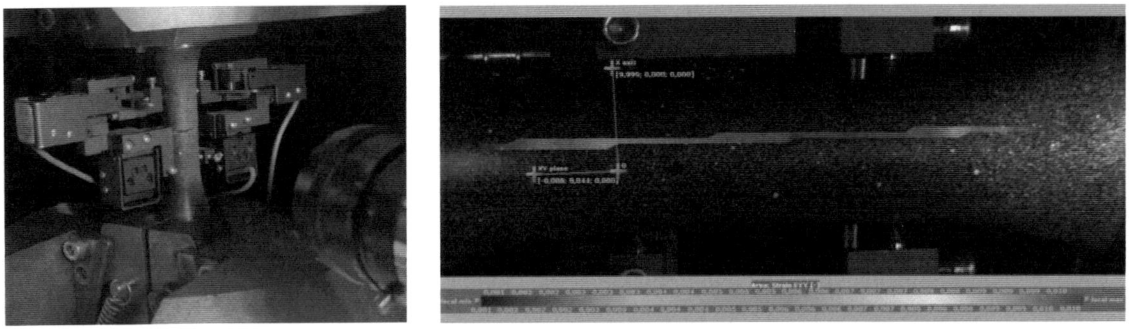

a) Experimental setup b) DIC strain contours corresponding to the strain peak

Fig. 3: Photos from DIC measurements focused on accelerated CSSC estimation.

The following steps make up the technique applicable in engineering practice:

1) In the preparatory DIC procedure, a special calibration grid containing coordinate axes is used. This grid is positioned using a special fixture, ensuring that the Y-axis is aligned with the axis of the test specimen. The software then detects the grid and sets the global coordinate system.
2) Determine the load amplitude $F_a$ at half-life from the records available from the fatigue testing machine.
3) In the DIC software, create an area probe (network of DIC points) where the axial strain will be computed.
4) The radius of the cross section corresponding to a given DIC point can be calculated from the Pythagorean equation:

$$r = \sqrt{X^2 + Z^2}, \tag{1}$$

where $X$ and $Z$ are the coordinates of the given DIC point, see Fig 4.

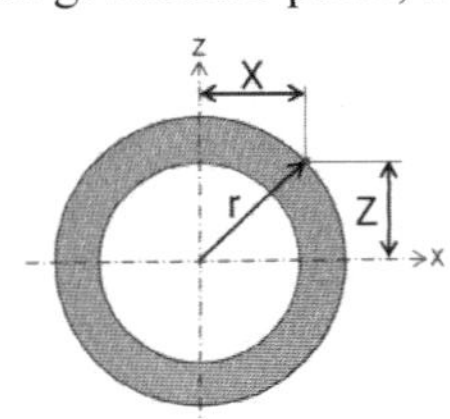


Fig. 4: Scheme of obtaining the cross-section radius from a DIC point.

5) The axial stress amplitude is calculated by

$$\sigma_a = \frac{F_a}{\pi r^2}. \tag{2}$$

6) Evaluate the amplitude of the axial strain for each DIC point.
7) Filter out the data corresponding to the straight part of the specimen and close to the border between rounded and testing part.
8) Use appropriate approximation function with the least-square method.

The previous procedure [3] was applied successfully also to get more ratcheting curves from a stress-controlled LCF test or more creep curves for an austenitic stainless steel [4].

## 5 Results

The experimental data of both accelerated techniques are compared with the conventionally obtained CSSC in Fig. 5. The DIC results were obtained by approximation of the strain amplitudes of both curved parts together (green points). Consideration of the DIC points from both curved parts of the specimen eliminates eventual parasitic bending due to misalignment of the grips. For validation, data from standard LCF tests conducted on specimens from the same build as the DIC specimens are also included (the approximation is in blue). The relative error between the CSSC obtained by the approximation curve of the DIC technique (black curve) and the approximation curve of the LCF tests is less than 3 %. Specimen from different SLM build was used in the case of the incremental fatigue test; thus, the agreement between the method proposed by Lukáš and Klesnil is not as good as that of the DIC technique.

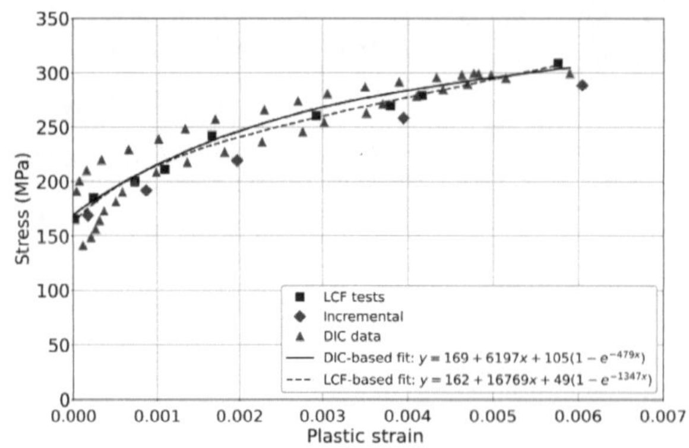


Fig. 5: Comparison of results from the two accelerated testing techniques with conventional CSSC.

## 6 Conclusion

In this study, the incremental fatigue test method was applied on a specimen from a different building, which explains the higher deviation from the conventional CSSC curve. Other results confirm the reliability of the DIC-based method for determining the cyclic stress-strain curve of additively manufactured AlSi10Mg, offering a relatively fast and accurate evaluation with lower time demands. A new procedure has been verified, which provides an easy evaluation of the CSSC curve. The most important innovation is the usage of the jig for placement of the calibration target into the position, where an axis of the coordinate system is aligned with the axis of the testing machine grips, i.e. the axis of the specimen after gripping. The validation of the new procedure in the cases of torsional and axial torsional, similarly as in [3], will be done as a future work.


## Acknowledgement

This work is an output of the project with financial support from the Czech Science Foundation under the registration no. 23-05338S and by the Specific Research Project SP2025/048.


## References


[1] Fini S. et al., Fatigue response of AlSi10Mg by laser powder bed fusion: influence of build orientation, heat, and surface treatments, Progress in Additive Manufacturing 10 (2025) 1385–1403, https://doi.org/10.1007/s40964-024-00712-y.

[2] Klesnil M., Lukáš P., Fatigue of Metallic Materials, Elsevier, 1992. ISBN: 9780444987366.

[3] Halama R. et al., A technique for estimation of cyclic stress-strain curves using an optical strain measurement and its application to additively manufactured AlSi10Mg, Procedia Structural Integrity 56 (2024) 111-119, https://doi.org/10.1016/j.prostr.2024.02.045.

[4] Halama R. et al., Cyclic plasticity of additively manufactured metals. In Cyclic Plasticity of Metals: Modeling Fundamentals and applications. Amsterdam: Elsevier, p.397–433, 2022. ISBN: 978-0-12-819293-1 https://doi.org/10.1016/B978-0-12-819293-1.00022-X

[5] Matušů M., Estimation of Service Life through Thermographic Methods. Doctoral thesis, Czech Technical University in Prague, Faculty of Mechanical Engineering, Department of Mechanics, Biomechanics and Mechatronics; Prague 2024. Available at CTU DSpace.

# X-ray Diffraction Analysis of Preferred Orientation and Residual Stresses of Aluminium Alloy Manufactured by Selective Laser Melting Technology

J. Čapek [1,*,a], K. Trojan [1,*], R. Halama [2], N. Ganev [1,*], J. Hajnyš [2]

[1] *Czech Technical University in Prague; Trojanova 13, 120 00 Prague, Czech Republic*
[2] *VSB - Technical University of Ostrava; 17. listopadu 2172/15, 708 00 Ostrava - Poruba, Czech Republic*
[*] *Center of competence "Smart mechatronic, eco-and energy-saving systems and technologies"*
[a] *jiri.capek@fjfi.cvut.cz*

**Abstract:** The paper deals with the influence of the crystallographic preferred orientation (texture) and residual stresses determined by X-ray diffraction on the fatigue life. It was observed that the texture is preserved in the building direction. Therefore, for the different inclined surface, different texture is presented. Based on texture and the RS, the theoretically relative fatigue life was estimated and subsequently correlated with the experimental fatigue life. Nevertheless, for the low-cycle fatigue test, texture and pores orientation in the 3D printed structure seem to be the main parameters affecting fatigue life.

**Keywords:** aluminium alloy; X-ray diffraction; preferred orientation; fatigue life; residual stresses.

## 1 Introduction

Additive manufacturing, in this case Selective Laser Melting (SLM) technology, is a promising metal powder consolidation method and offers outstanding parts production opportunities. The complexity of the SLM process results in the residual stresses (RS) being highly dependent on laser power, scanning speed, scanning strategy, and other processing parameters [1]. During the SLM process, various RS are generated due to the inhomogeneous printing process and cooling.

The initiation and growth of fatigue cracks are critical factors that influence fatigue behaviour and are closely related not only to the RS values, but it is a complex problem, where surface roughness, microstructural characteristics (such as dislocation density, crystallite size, and microstrains), pores, and their orientation influence the crack formation [2]. The crystallographic preferred orientation (texture), as a result of samples' anisotropy, has a direct influence on the mechanical properties [3]. For face-centered cubic (fcc) materials, such as iron (γ-Fe), nickel, and aluminium, the *[111]* crystallographic orientation have maximum value of Young modulus and the *[001]* the minimum value. For yield strength, the highest and lowest values for fcc materials appear in the *[111]* and *[110]* crystallographic orientations, respectively [4, 5].

The previous study [2] showed that the state of the surface residual stress is not homogeneous. These so-called surface microstructure notches (step changes in RS) that can be prone to fatigue crack formation and can have a major influence on fatigue life were found. Nevertheless, not only the surface RS are important from a fatigue life point of view, but also their subsurface or bulk values are convenient to know. These stresses can also distort the geometry of the printed object, with some shapes, especially those created using a horizontal printing strategy, being more susceptible than others. High tensile RS, in particular, can cause deformation and cracking, thereby compromising whole structural integrity. In general, tensile RS within the surface and subsurface layers can deteriorate both mechanical and corrosion properties.

It is generally known that the yield strength significantly influences a material's fatigue life, but the relationship is not always straightforward. Generally, materials with higher yield strengths tend to exhibit better fatigue resistance, especially at lower stress levels. Young's modulus can influence fatigue life in several ways. Generally, a higher Young's modulus (stiffer material) can lead to a lower fatigue life, as it may be more susceptible to brittle fracture under cyclic loading. However, the relationships are complex, and other factors like material microstructure, surface, loading conditions, etc. also can play a significant role.

Therefore, the impact of orientation of printed samples on the texture and RS of as-built bulk material was investigated at several angles of the sample axis to the printing platform using X-ray diffraction.

## 2 Methodology

For the fabrication of AlSi10Mg aluminium alloy tubular samples with 14 mm outer diameter and 11.35 mm inner diameter from recycled powder (15× sieved) with the chemical composition of 9–10 wt.% Si, 0.2–0.45 % Mg, ≤0.55 % Fe, ≤0.1 % (Cu, Mn, Zn, Ni, Pn, Sn each), ≤0.15 % Ti, and Al (bal.). Renishaw AM500S was used with the printing parameters shown in the Tab. 1. Six samples were printed with different inclined angles of 0°, 10°, 20°, 30°, 40°, and 90° (slope related to the normal direction of printing platform). Two directions will be used in the text: AD (axial direction of the sample) and TD (transversal direction, perpendicular to the axial).

Tab. 1: Printing parameters.

| parameter | |
|---|---|
| laser power [W] | 350 |
| spot size [μm] | 70 |
| layer thickness [μm] | 30 |
| printing strategy | meander |
| laser speed [mm/s] | 1150 |

The Empyrean diffractometer with cobalt radiation was used for the texture (preferred orientation) analyses by X-ray diffraction (XRD). Texture analysis was performed based on the calculation of the orientation distribution function (ODF) from experimental pole figures which were obtained on three diffraction lines *{311}*, *{222}*, and *{400}* of aluminium phase. The MATLAB™ toolbox MTEX software [6] was used for processing the data and plotting of inverse pole figures (IPF) and ODF.

Macroscopic residual stresses (RS) were determined using the X'Pert Pro MPD diffractometer by XRD when the *{311}* diffraction line of the aluminium phase was measured using chromium radiation in the direction parallel to the rotation axis of the sample. The diffraction angles $2\theta^{hkl}$ were determined from the diffraction lines $K\alpha_1$, fitted using the *Pearson VII* function, and Rachinger's method was used for the separation of the diffraction lines $K\alpha_1$ and $K\alpha_2$. For RS determination, the $\sin^2\psi$ method and X-ray elastic constants $\frac{1}{2}s_2 = 19.05$ TPa$^{-1}$ and $s_1 = -4.89$ TPa$^{-1}$ were used. The experimental errors given for each value are the standard deviations of the calculation of the RS [7].

The XRD measurements were performed on the as-built bulk material at the same vertical distance of 15 mm from the printing platform in the grip section of the samples.

## 3 Results and Discussions

From the results of the texture analysis, it can be observed that the texture is preserved in the building direction (that is, in the normal direction to the printing platform). This implies that the mechanical properties will be different for each inclined sample in the given direction; therefore, all results will be presented from the samples orientation point of view. From the IPFs (Fig. 1), it can be seen that the 0–20° and 90° samples have very similar texture in the AD, therefore similar mechanical properties can be expected. For the 30° and 40° samples, there is already a significant rotation of the texture, so the properties can be different; see Fig. 1. Moreover, the strength of the texture (m.r.d. units; multiple of random distribution), or anisotropy, is higher for 0° sample (2.5× higher in comparison with non-textured material).

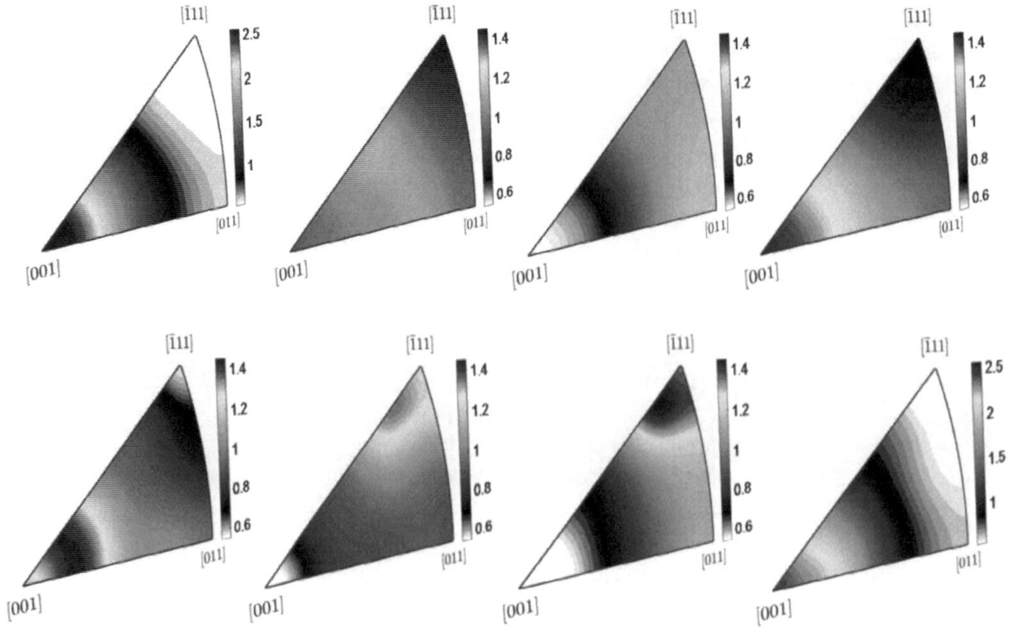


a) 0° sample b) 30° sample d) 40° sample e) 90° sample

Fig. 1: Inverse pole figures in the axial direction (top) and transversal direction (bottom).

Nevertheless, the IPFs do not contain all information about texture; therefore, the ODF has to be discussed. In Fig. 2, there are typical sections for fcc materials, where typical texture fibres and components are visible. For sample 0°, the η-fibre <*100*>‖AD with two main texture components (cubic and Goss). This proves the IPF result and for 0° sample, the Young's modulus should be smaller in comparison with other samples. Sample 10° has the same texture fibre but with smaller m.r.d. values, and the maximum of cubic texture component was shifted about 10°. After other rotations, i.e. 30° and 40° samples, new texture components were observed. All these texture components have approx. direction *{011}*<*110*>, which theoretically results in decreasing of the yield strength. The 90° rotation of the origin texture results in ε-fibre <*100*>‖TD with the cubic and rotation cubic component which is not evident from the IPFs.

Residual stress is other parameter that has a direct influence on the fatigue life of the material. From this point of view, see Fig. 2, the fatigue life should be lowest for the 0° and 10° samples where the highest undesirable tensile bulk and surface [2] RS were found.

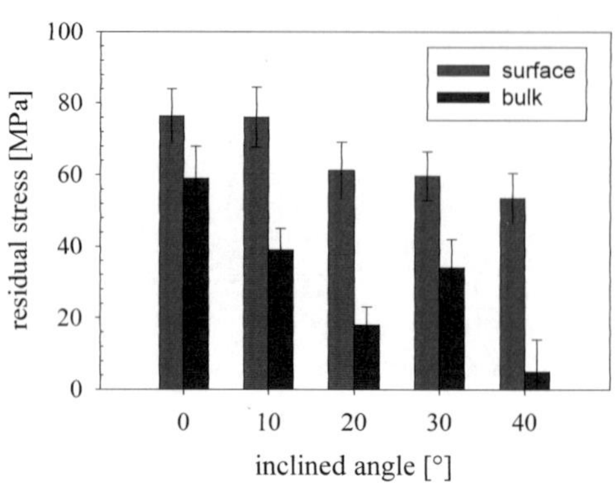


Fig. 2: Residual stresses with respect to different inclined samples in the axial direction.

# 4 Conclusion

According to the texture, the following results can be theoretically predicted:

- the fatigue life of 0° sample should be the highest, due to theoretically low Young's modulus,
- the fatigue life of 40° sample should be the worst, due to theoretically high Young's modulus and low yield strength.

Nevertheless, only from RS point of view, the fatigue life should be opposite, in comparison with texture influence.

After the mechanical tests, the highest and the lowest yield strength were found 0° and 10° samples, respectively. Average value of yield strength was observed for 40° sample. The Young's modulus is approx. same within the experimental error for all samples.

From the low-cycle fatigue test (results will be the subject of another article in processing) follows that the smallest fatigue life was found for the 0° sample due to the presence of pores in the critical layer parallel to the printing platform and probably because of high tensile RS. The decrease in fatigue life was proven for the 40° sample, due to the combination of the unfavorable texture (where the *{011}<110>* texture component dominates), the presence of microstructural notches on the surface, see [2], and the presence of the pores in the shear direction.

In conclusion, it can be noted that the influence of the RS was reduced, probably due to the high applied strain amplitude during fatigue tests. Therefore, the overall mechanical properties as well as the fatigue life are a set of many influences and parameters, not only texture and RS. It seems that the way of testing also affects the effectiveness of each parameter. For the low-cycle fatigue test, texture and pores orientation in the 3D printed structure seem to be the main parameters affecting fatigue life.

# Acknowledgement

The work on this paper was funded by the Czech Science Foundation grant 25-15579S. This research was supported by the European Regional Development Fund within the OP "Research, Innovation and Digitalization Programme for Intelligent Transformation 2021-2027", Project No. BG16RFPR002-1.014-0005 Center of competence "Smart Mechatronics, Eco- and Energy Saving Systems and Technologies" and the Grant Agency of the Czech Technical University in Prague, grant No. SGS25/168/OHK4/3T/14.

# References

[1] J. L. Bartlett, X. Li, An overview of residual stresses in metal powder bed fusion, Additive Manufacturing 27 (2019) 131–149, http://dx.doi.org/10.1016/j.addma.2019.02.020.

[2] J. Čapek et al., Residual Stress Analysis of Additively Manufactured AlSi10Mg Alloy, in proc.: Proceedings of Experimental Stress Analysis (2023), ed. M. Hagara, R. Huňady, Technical University of Košice, Košice, 22–26.

[3] L. Ya. Volozneva, P. G. Miklyaev, Anisotropy of the strength characteristics of aluminum alloys, Noferrous Metals and Alloys 12 (1970) 586–590, https://doi.org/10.1007/BF00664954.

[4] I. L. Svetov et al., Anisotropy of mechanical properties of single crystals in fourth generation Ni-based superalloy, in proc.: Materials for advanced power engineering 2010 (2010), ed. J. Lecomte-Beckers et al., Liege, 660–667.

[5] A. M. Korsunsky, M. R. Daymond, K. E James, The correlation between plastic strain and anisotropy strain in aluminium alloy polycrystals, Materials Science and Engineering: A 334 (2002) 41–48, https://doi.org/10.1016/S0921-5093(01)01780-4.

[6] Retrieved from: http://www. https://mtex-toolbox.github.io/

[7] I. Kraus, N. Ganev, Technické aplikace difrakční analýzy, Praha, ČVUT, 2004.

# Design of a Sensor Network for In-flight Thrust Measurements of a Jet Training Aircraft Propulsion Unit

R. Doubrava [1,a], J. Raška [1]

[1] VZLU AEROSPACE, Czech Republic

[a] doubrava@vzlu.cz

**Abstract:** Accurate in-flight measurement of propulsion and payload forces is essential for optimizing aircraft performance, yet conventional certification methods often lack the ability to capture such data. This study presents the development of a smart sensor network based on strain gauges, designed to directly measure engine thrust and external loads on rigid aircraft structures. Two approaches were proposed: expanded strain gauge networks for eliminating secondary load effects, and concentrated systems that reconstruct structural elements into dedicated load-measuring components. A multi-level numerical simulation strategy was applied, beginning with simplified 2D finite element analyses and advancing to high-fidelity 3D models. These simulations guided optimal placement and wiring of strain gauges to ensure sensitivity, minimize cross-load interference, and maintain linear correlation with applied forces. To validate the design, a dedicated test stand with a dummy engine was developed for calibration under controlled static loading. The validated system enables robust, real-time thrust monitoring in flight.

**Keywords:** flight measurement; engine thrust; sensor; FEM; test.

## 1 Introduction

Increasing aircraft performance is closely linked to obtaining real data during flight measurements. To a large extent, recommended or standardized certification procedures are used in these measurements. However, these procedures typically do not provide additional operational data that cannot already be obtained from simulations or wind tunnel testing [1]. Parameters such as engine thrust or payload resistance are essential for performance optimization but are often not directly accessible through conventional certification methods.

Measurement of force quantities in aerospace applications is generally based on detecting deformations in elastic structures. In aircraft, particularly in areas such as engine mounts or payload attachment points, the structural components are relatively rigid, which reduces the sensitivity of standard resistance strain gauges [2]. Moreover, the complexity of in-flight loadings such as vertical and lateral gusts, gyroscopic moments, vibrations, and temperature variations further complicates accurate force measurement [3].

To overcome these limitations, advanced methods have been developed, including direct load measurement on engine mounting systems using strain gauges, which allow for independent thrust estimation without relying on engine or propeller manufacturer data [2]. Guidelines and methodologies for such in-flight thrust determination are outlined in standards such as [1] and [4], which provide comprehensive frameworks for planning, instrumentation, and error analysis in thrust measurement campaigns.

## 2 Material and methods

The proposed smart sensor systems, based on standard strain gauge sensors, will have the following characteristics:

- Expanded strain gauge networks - strain gauges will be distributed on the structure in such a way that the effects of secondary load components are mutually eliminated already in the connection of the own strain gauge network so that the resulting signal from the network will represent a response to the required component of external loads.

- Concentrated strain gauge system - the structural element will be reconstructed in the style of a strain gauge so that the individual strain gauges will directly measure the components of the external load in the appropriate places of the strain gauge element in the direction of the axes of the selected coordinate system.

## 2.1 Numerical simulation

A network of strain gauge sensors placed on the real structure (engine attachment) was designed for the implementation of the measurement. Numerical models will be used to optimize the design of the network of measuring sensors, including the effects of real boundary conditions affecting the actual measurements. The design of the measuring elements will be experimentally verified on test stands under static loading simulating real operation. The optimized measuring network and measuring members will be designed to eliminate secondary loads and allow direct measurement of required load values, such as engine thrust or resistance of suspended loads, with the possibility of a follow-up to adjust flight characteristics in real time.

Based on the preliminary FEA of the engine mount (2D plate elements), the strain gages' position and wiring system were proposed, in order to sensitively measure the motor thrust and eliminate the parasitic (vertical and lateral) forces (Fig.2).

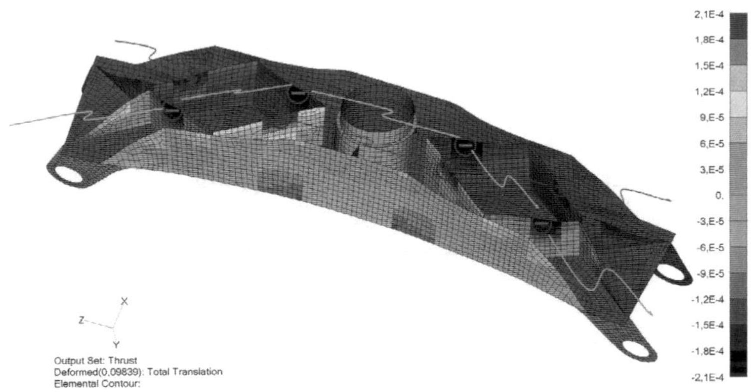


Fig. 2: FEA preliminary analysis, strain gage positions, and wiring system.

To refine and validate this preliminary design, a detailed 3D finite element model of the engine mount was subsequently developed (Fig. 3). This model incorporates the real geometry of the structure, including fillets, radii, thickness transitions, and other geometric discontinuities that significantly influence local stress and strain distributions. The 3D model allows for a more accurate representation of stress concentrations and load paths, which are critical for the correct placement and orientation of strain gauges.

The simulation results from the 3D model were used to:

- Verify the effectiveness of the proposed sensor layout under realistic loading conditions.
- Identify potential areas of stress concentration that could affect measurement accuracy or sensor durability.
- Adjust the sensor network to account for local geometric effects that were not captured in the simplified 2D model.
- Ensure that the measured strain signals correlate linearly with the applied thrust force, even in the presence of complex multi-axial loading.

This multi-level modeling approach—starting with a simplified 2D analysis and followed by a high-fidelity 3D verification—ensures both computational efficiency during the design phase and high accuracy in the final sensor network configuration.

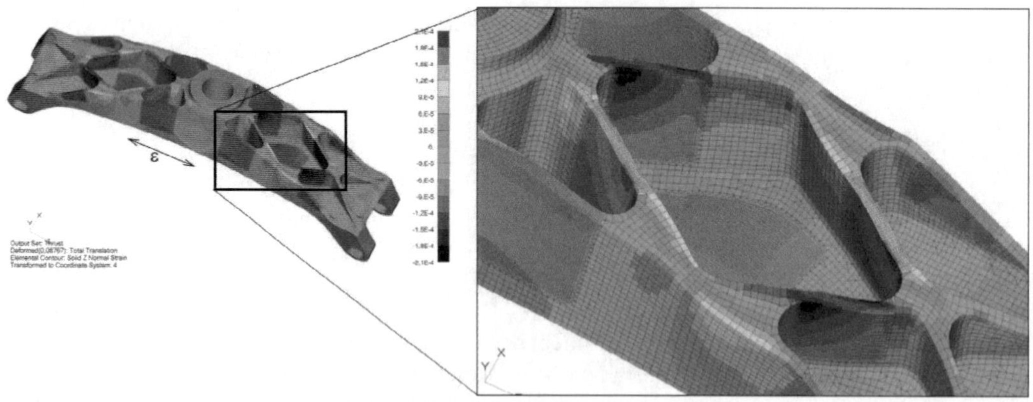

Fig. 3: Detailed FE analysis.

# 3 Test and calibration

To validate the functionality and accuracy of the optimized measuring network, a dedicated test stand was developed. This stand is designed to simulate the static loading conditions that occur during real aircraft operation, particularly those acting on the engine mount structure. The test stand enables controlled application of axial, lateral, and vertical loads, allowing for precise calibration and verification of the sensor system under known boundary conditions (Fig.4).

A key component of the test setup is a dummy engine unit (Fig.5), which replicates the mass, geometry, and mounting interface of the actual propulsion unit. This surrogate allows for safe and repeatable testing without the risks associated with live engine operation. The dummy is equipped with integrated load application points and interfaces for mounting the strain gauge network and associated electronics [5].

Fig. 4: CAD and real calibration stand.

Fig. 5: Engine dummy for experimental verification of sensor network design to measure engine thrust in flight.

The calibration process includes the following steps:

- Installation of the measuring elements (strain gauges and wiring) on the engine mount structure and dummy engine.
- Application of known loads using hydraulic or mechanical actuators in multiple directions to simulate thrust, side loads, and vertical forces.
- Data acquisition and signal processing, where the output of the strain gauge network is recorded and compared with reference load cell measurements.
- Adjustment and tuning of the sensor configuration to ensure linearity, repeatability, and minimal cross-sensitivity to parasitic loads.

This experimental phase is essential for:

- Verifying the theoretical assumptions made during the numerical simulations (2D and 3D FEA).
- Identifying any nonlinearities or unexpected interactions in the real structure.
- Establishing calibration curves and correction factors for in-flight data interpretation.
- Ensuring mechanical robustness and signal stability of the measuring elements under operational conditions.

The validated sensor system will then be ready for integration into the aircraft for real-time thrust monitoring.

# 4 Conclusion

A network of strain gauge sensors placed on a real structure (motor mounting) was designed to implement measurements. Numerical models were used to optimize the design of the network of measuring sensors, including the effects of real boundary conditions affecting the measurements themselves. The design of the measuring elements will be experimentally verified on test benches under static loads simulating real operation.

# Acknowledgement

The result was realized as part of the project solution number TN02000009 National Competence Center for Aviation and Cosmonautics, sub-project TN02000009/03 "Smart, Safe and Efficient Aircraft".

# References

[1] SAE AIR1703A: In-Flight Thrust Determination, 2017

[2] H. Muhammad et al., IN-FLIGHT THRUST ETERMINATION BY LOAD MEASUREMENT ON THE ENGINE MOUNTING SYSTEM, ICAS 2000

[3] Jebáček, I., Horák, M. (2012). Possibilities and Methods of In-Flight Loading Measurement, Brno University of Technology, AVIATION Volume 16(2): 47–50, doi:10.3846/16487788.2012.701860, 2012

[4] AGARD No. 237: Guide to In-Flight Thrust Measurement of Turbojets and Fan Engines, ISBN 92-835-1304-5, 1979

[5] Novotný, D et al., METODIKA KALIBRACE TENZOMETRICKÉ SÍTĚ PRO MĚŘENÍ TAHU PROUDOVÉHO MOTORU, zpráva VZLÚ R-7678, 2024.

# Experimental Study of Inclined Angle Influence on Static and LCF properties of Additively Manufactured Aluminium Alloy

R. Halama [1,a], M. Kořínek [1], J. Čapek [2], K. Trojan [2], N. Ganev [2], J. Hajnyš [1]

[1] *VŠB - Technical University of Ostrava, Ostrava, Czech Republic*
[2] *Czech Technical University in Prague, Trojanova 13, 120 00 Prague, Czech Republic*
[a] *radim.halama@vsb.cz*

**Abstract:** The influence of orientation on mechanical properties of additively manufactured material with respect to anisotropy and fatigue is usually investigated under three angles of the specimen axis to the building platform – vertical, horizontal, and 45°. The reason is that there is a rule of thumb in most additive technologies regarding the addition of supports when the tangent plane to the platform inclines the angle below 45°. If this rule is not followed, the roughness of the surface will be poor. Surprisingly, there is a lack of studies investigating differences in fatigue properties for angles different than 45°. This study focusses on the tensile and low-cycle properties of AlSi10Mg prepared by selective laser melting. The test specimens were printed for 0°, 10°, 20°, 30°, and 40° of inclined angle. The fatigue test results are supported by X-ray diffraction measurements and the yield probing technique.

**Keywords:** Additive manufacturing; low-cycle fatigue; X-ray diffraction; Yield surface tracing; anisotropy.

## 1 Introduction

Additive manufacturing (AM) using laser powder bed fusion (LPBF) enables the production of complex metallic components with high geometrical flexibility. However, the layered nature of this process leads to pronounced anisotropy in mechanical properties, particularly in terms of fatigue behaviour. Most studies focus on three standard orientations of the specimens with respect to the building platform: horizontal, vertical, and inclined (45°). These orientations reflect common practice in AM design, especially due to the rule of thumb that support structures are required when the angle of inclination measured from the platform is below 45°, otherwise the surface quality significantly deteriorates.

Although the influence of these three orientations on static and fatigue performance has been partially explored, there is a lack of systematic studies addressing mechanical behaviour of specimens manufactured at angles below 45° (considering definition in Fig. 1). This angle range is critical in terms of both the necessity of the support structure and surface quality. For example, Fini et al. (2025) investigated the fatigue properties of AlSi10Mg in three orientations (0°, 45°, 90°) and found a significant effect of the build direction on fatigue strength, with improved performance after heat treatment with T6 [1]. Similarly, a study published in the Rapid Prototyping Journal (2021) reported that orientations ranging from 0° to 75° influence porosity and tensile strength, with lower angles showing higher porosity and reduced performance [2].

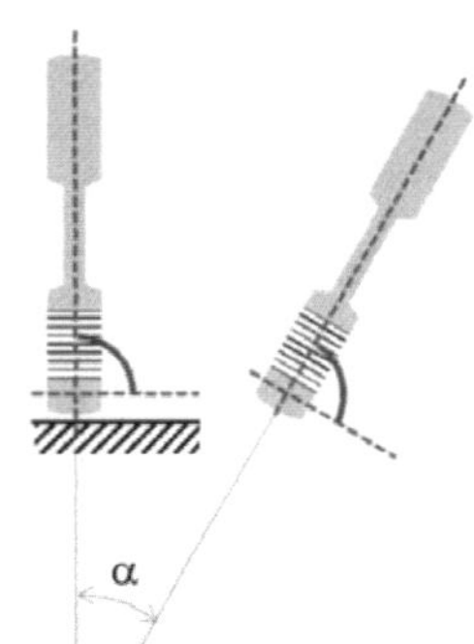


Fig. 1: Inclined angle definition.

Given these findings, a detailed investigation of the mechanical properties of specimens printed at low inclination angles is essential to optimize LPBF component design and performance. The present study focuses on the tensile and low-cycle fatigue (LCF) behaviour of AlSi10Mg fabricated using a Renishaw AM500E printer. The specimens were manufactured at inclination angles of 0°, 10°, 20°, 30°, and 40° and tested under strain-controlled conditions. The goal is to provide a comprehensive understanding of how small changes in the orientation of the construction influence the mechanical response, an area that remains largely underexplored.

# 2 Material

The AlSi10Mg alloy represents one of the most widely utilized materials in lightweight structures fabricated through LPBF and is frequently employed within the aerospace [3] and automotive sectors owing to its exceptional resistance and fatigue characteristics [4].

## 2.1 Additive technology

For the purpose of fabricating specimens, the three-dimensional printer manufactured by Renishaw, specifically the AM500S model, was used. The specimens were produced utilizing a laser power of 350 W, a scanning velocity of 1150 mm·s$^{-1}$, and a layer thickness measuring 30 μm. A scanning strategy referred to as "Meander" (as illustrated in Fig. 2) was implemented, which is recognized as the predominant strategy for specimens characterized by minimal cross-sectional areas. Throughout the fabrication process, the chamber was purged with argon gas of purity 5.0 to prevent the occurrence of oxidation. The apparatus was subjected to a preheating process at a temperature of 170 °C in order to mitigate the residual stresses inherent in the material.

Fig. 2: Scanning strategy "Meander"; Reproduced with permission from [5].

## 2.2 Testing equipment and conditions

In our previous study [6, 7], basic LCF properties were reported and the influence of mean stress and stress amplitude on the uniaxial ratcheting behaviour of the AlSi10Mg aluminium alloy was shown in incremental testing. This time, we focus only on strain-controlled cases. All tests were performed at room temperature on as-built specimens without any subsequent heat treatment.

### 2.2.1 Tensile testing

Tensile testing was conducted using the LabControl 20kN testing machine with the help of digital image correlation (DIC). Two 2.3Mpx CCD cameras with 20 FPS were applied. Alpha software developed by the X-Sight company was used in measurement and evaluations. The position rate of 10 mm/min was applied based on previous observations that the aluminium alloy is not sensitive to the strain rate.

### 2.2.2 Fatigue testing

All LCF tests were performed using the LabControl 100kN/1000Nm hydraulic fatigue testing machine. The tests were carried out under total strain amplitude control, applying a constant strain rate of 1.0 %/s. The strain amplitudes considered for all inclined angles in this study are 0.65 %, 1.0 %, and 1.5 %. For vertically printed specimens, additional strain amplitudes (0.87 % and 1.1 %) were considered. The Model 3452 extensometer manufactured by Epsilon Technology Corp. with a gauge length of 10 mm was applied to measure the longitudinal strain in the case of strain amplitudes less than 1.2 %, while the solid specimen with 6.35 mm diameter was used. In the case of higher strain amplitudes, a thin-walled specimen was designed with an outer diameter of 14 mm and an inner diameter of 11.5 mm connected with the application of the Epsilon Technology Corp. 3550 extensometer model.

# 3 Results

The results obtained from the mechanical testing of the AlSi10Mg alloy manufactured through LPBF are presented in two separate sections, dedicated to the tensile tests, yield surface probing tests, and the LCF tests.

## 3.1 Tensile test results

The resulting engineering stress-engineering strain curves are shown in Fig. 3a. The lowest ductility value was obtained, together with the highest yield and the ultimate strengths, for the vertically printed specimen; see Fig. 3bcd. The highest ductility corresponds to the smallest inclination of the specimen's axis (10°), leading to the expectation of the best lifetime in the same case.

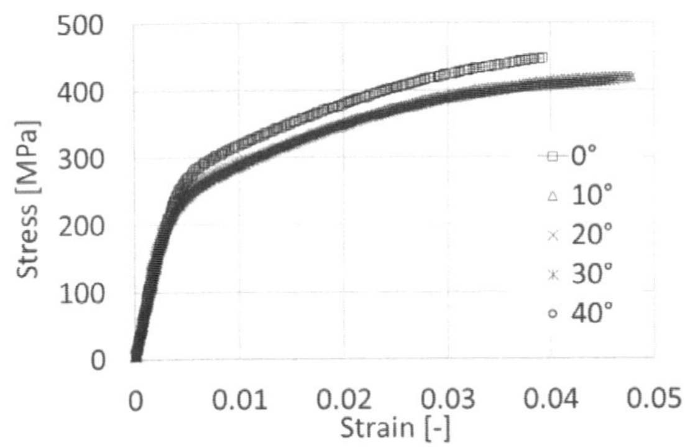


a) Engineering stress-strain curves

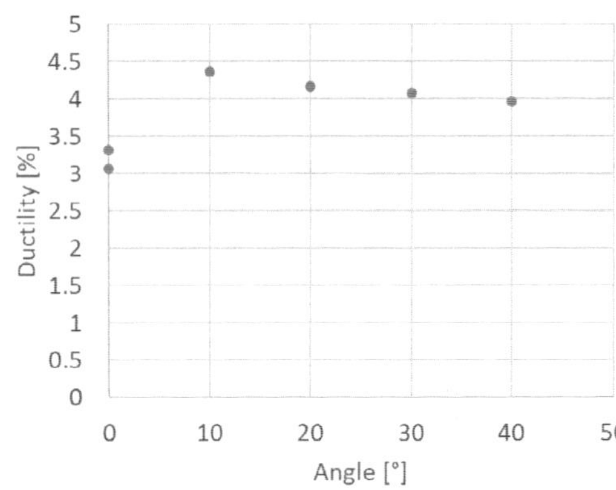


b) Ductility as a function of the inclined angle

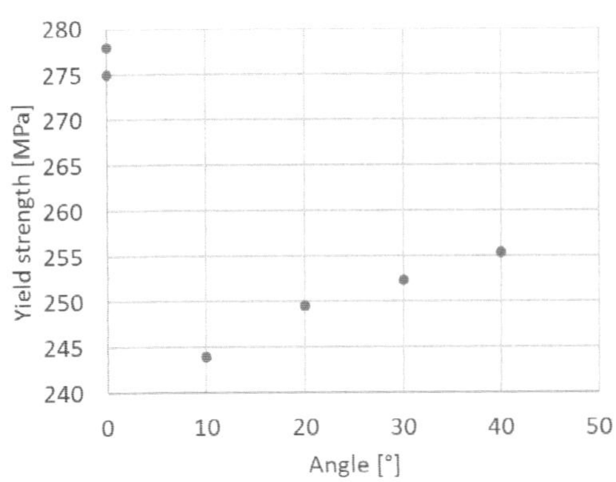


c) Yield strength as a function of the inclined angle

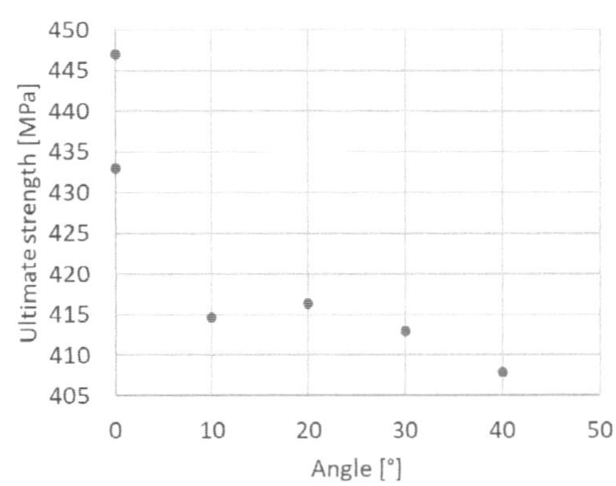


d) Ultimate strength as a function of the inclined angle

Fig. 3: Results of the tensile tests.

## 3.2 Yield surface tracing results

Tracing of the yield surfaces was also performed to observe the effect of the printing direction on the size and shape of the elastic region. The experiment was carried out as described in [8], including test equipment. A yield criterion based on the increment of plastic strain (also called offset strain) defined by Eq. (1) was used to evaluate the specific yield stress (points in Fig. 4ab).

$$\Delta\varepsilon_{eff} = \sqrt{\Delta\varepsilon_{pl}^2 + \frac{1}{3}\Delta\gamma_{pl}^2}, \quad (1)$$

where $\Delta\varepsilon_{pl}$ is the increment of plastic strain in the axial direction and $\Delta\gamma_{pl}$ is the increment of plastic shear strain.

The measurements were repeated for both specimens a total of 3 times. The values of each specific yield stress were then averaged. Figure 4a shows the yield surfaces for both specimens with an offset strain value of 50 μS. Fig. 4b shows the yield surfaces with an offset strain value of 100 μS. The measured values could not be described by a circle, the von Mises condition, but were interleaved with an ellipse. It can be seen that for both specimens the shear stress component shows values higher than those in the axial direction of the specimens. Similarly, the centre of the yield surface is not directly at the origin of the coordinate system but is slightly displaced into the compressive part. Overall, the printing inclination does not have a significant effect on the size and shape of the initial plasticity surface, as can be seen in the figures. The most significant differences are in the pressure part of the surface, with higher values of the specific yield stress for the specimen produced at an angle of 40°. In that case, only slight differences in the semiaxis of the ellipse are visible (Fig. 4b).

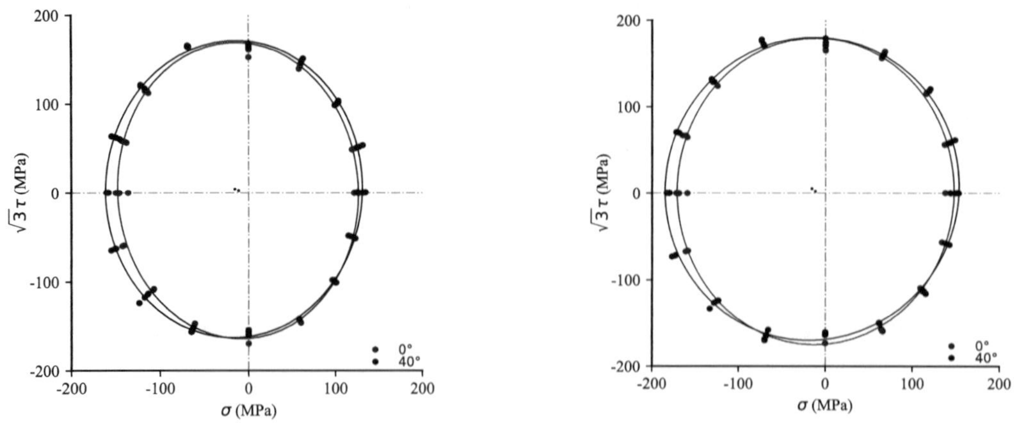


a) Offset strain 50 μS b) Offset strain 100 μS

Fig. 4: Comparison of yield surface shape for vertically printed specimen and the inclined by 40° angle [9].

## 3.3 Fatigue test results

As visible in Fig. 5b, at 0.65 % and 1 % of the axial strain amplitude, the lifetimes are approximately 20 % higher for 10° than in the case of a vertically printed specimen, which is in good correlation with the ductility trend observed in tensile tests. At the highest level of axial strain (1.5 %), the differences are not so noticeable. However, the specimens used for the 1.5 % strain amplitude were also previously used to identify the eventual anisotropy using the yield tracking technique [9], so the lifetimes could be affected. Furthermore, the 10° specimen shows a significantly higher value of the stress amplitude, as can be seen in Fig. 5a, which logically leads to fewer cycles to failure.

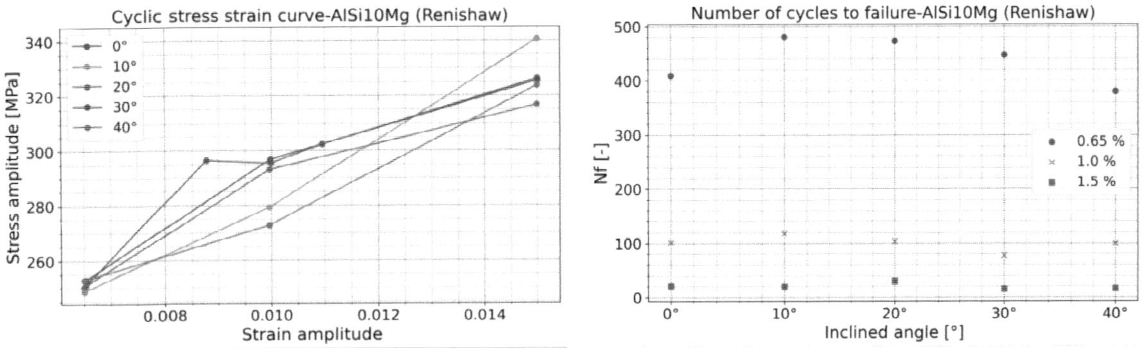


a) Cyclic stress-strain curve comparison b) Lifetimes as a function of the inclined angle

Fig. 5: Results of fatigue tests performed on AlSi10Mg fabricated additively.

Although it has the highest yield stress for 0°, the stress amplitude after saturation is almost the same as for 20° and 30° (Fig. 5a). Plastic work evaluated in the half-life defined by Eq. (2) is the commonly applied fatigue parameter for aluminium alloys.

$$W_p = \int_{(cycle)} \sigma d\varepsilon_{pl}, \tag{2}$$

where $\sigma$ is the axial stress and $d\varepsilon_{pl}$ is the increment of axial plastic strain. The lifetime curve presented in Fig. 6 clearly shows that the fatigue resistances for 10° and 20° are comparable and the best ones.

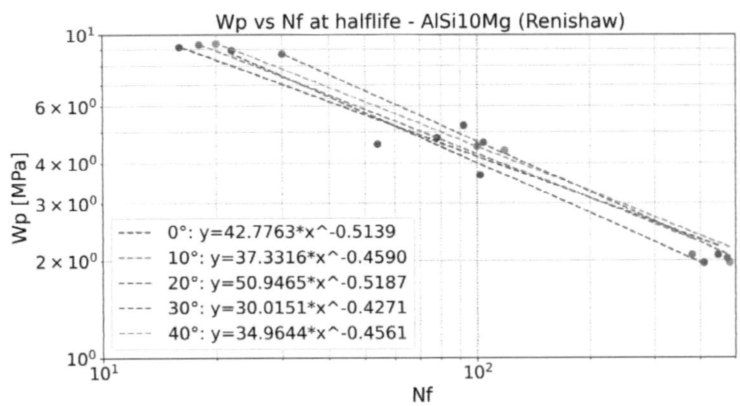


Fig. 6: Plastic work as a function of number of cycles Nf to failure.

# 4 Conclusion

The results of this study demonstrate that even small deviations from the vertical build orientation can have a significant effect on the mechanical behaviour of additively manufactured specimens. In this study, the inclination angles of 0°, 10°, 20°, 30°, and 40° were considered. The 10° inclined specimens exhibited the highest ductility and improved fatigue life at moderate strain amplitudes. Vertical specimens showed the highest strength but the lowest ductility, and their distinct stress–strain response suggests a different deformation mechanism. The worst fatigue life is for 0° or 40°. The result seems to be more apparent for lower strain amplitudes, that motivates us to extend the study by fatigue testing on lower amplitudes.

The results of LCF tests for specimen 40° correspond to the results of the crystallographic preferred orientation analysis using XRD (the results are the subject of another article) where they are proven due to the unfavourable texture (where the [110] crystallographic orientation with the lowest yield strength dominates in the loading direction) for higher inclined angles of the specimen. The anisotropy was proved by the yield probing technique. Even a small inclined angle (10° or 20°) can extend the fatigue life of a component made from AlSi10Mg. These findings emphasise the importance of considering a wider range of build angles when optimising AM component performance for fatigue-critical applications.

## Acknowledgement

This work is an output of the project with financial support from the Czech Science Foundation under the registration no. 25-15579S (Experimental and numerical research on relaxation of residual stresses induced by selective laser melting) and by the Specific Research Project SP2025/048.

## References

[1] S. Fini et al., Fatigue response of AlSi10Mg by laser powder bed fusion: influence of build orientation, heat, and surface treatments, Progress in Additive Manufacturing 10 (2025) 1385–1403, https://doi.org/10.1007/s40964-024-00712-y.

[2] G. Tarakçı et al., Effect of building orientations and heat treatments on AlSi10Mg alloy fabricated by selective laser melting: microstructure evolution, mechanical properties, fracture mechanism and corrosion behavior, Rapid Prototyping Journal 28 (2022) 1609–1621, https://doi.org/10.1108/RPJ-11-2021-0325.

[3] A. Raja et al., A review on the fatigue behaviour of AlSi10Mg alloy fabricated using laser powder bed fusion technique, Journal of Materials Research and Technology 17 (2022) 1013–1029, https://doi.org/10.1016/j.jmrt.2022.01.028.

[4] N.T. Aboulkhair et al., Improving the fatigue behaviour of a selectively laser melted aluminium alloy: Influence of heat treatment and surface quality, Materials & Design 104 (2016) 174–182, https://doi.org/10.1016/j.matdes.2016.05.041.

[5] J. Mesicek et al., Abrasive Surface Finishing on SLM 316L Parts Fabricated with Recycled Powder. Applied Sciences 11 (2021) 2869, https://doi.org/10.3390/app11062869.

[6] F. Sari, R. Halama, J. Hajnyš, On Uniaxial Ratcheting of AlSi10Mg Prepared by SLM Technology, in proc.: Proceedings of Experimental Stress Analysis (2022), ed. Doubrava K., Kubasova K., Padovec Z., Blaha D., Sedlacek R. & Ruzicka M., CTU in Prague, Czechia, Prague, 353–359.

[7] P. Das et al., Constitutive modelling and damage prediction of AlSi10Mg alloy manufactured by SLM technology with emphasis on ratcheting in LCF regime, International Journal of Fatigue 181 (2024) 108115, https://doi.org/10.1016/j.ijfatigue.2023.108115.

[8] M. Kořínek et al., Determination of the Initial Anisotropy of Inconel 718 manufactured by Additive Manufacturing, in proc.: Proceedings of Experimental Stress Analysis (2023), ed. M. Hagara & R. Huňady, TU Košice, Slovakia, Košice, 102–106.

[9] M. Kořínek. Experimental Determination of Yield Surface Evolution under Cyclic Loading. Ostrava, 2024, Dissertation. Ostrava: VSB – Technical University of Ostrava, Faculty of Mechanical Engineering, Department of Applied Mechanics. 176 p. Thesis supervisor: prof. Ing. Radim Halama, Ph.D. (in Czech)

# Predicting the Fatigue Lifetime of Laser Powder Bed Fusion-processed Alloys Using Physics-informed Neural Networks

J. Halamka[1,*], O. Stránský[2], L. Beránek[2], M. Bartošák[1]

[1] *Department of Mechanics, Biomechanics and Mechatronics, Faculty of Mechanical Engineering, Czech Technical University in Prague, Technicka Street 4, 16607, Prague 6, Czech Republic*

[2] *Department of Machining, Process Planning and Metrology, Faculty of Mechanical Engineering, Czech Technical University in Prague, Technicka Street 4, 16607, Prague 6, Czech Republic*

* *jiri.halamka@fs.cvut.cz*

**Abstract:** In this work, conventional fatigue life models are coupled with a physics-informed neural network that incorporates defect characteristics obtained via X-ray computed tomography. The neural network is designed with modifications to incorporate known physical principles, ensuring physically meaningful predictions. The resulting hybrid model is trained and validated on experimental datasets obtained for laser powder bed fusion-processed 316L steel and AlSi10Mg aluminium alloy. The hybrid model shows significantly improved prediction accuracy compared to conventional models that do not account for defects.

**Keywords:** Physics-informed neural networks; Low-cycle fatigue; Fatigue lifetime prediction; Additive manufacturing.

## 1 Introduction

The damaging effect of multiple deposition defects on the low-cycle fatigue life of additively manufactured alloys remains insufficiently understood. While these defects must be considered collectively, their individual and combined effects on fatigue life are not well characterized. A machine learning approach presents a promising solution. This work builds on insights from our previous studies [1] and [2], which explored the interaction between defect morphology and fatigue behaviour. By incorporating defect characterization through machine learning models, we aim to develop a unified framework that captures the influence of complex defect features on fatigue life across various materials and loading conditions.

Section 2 provides a brief overview of Artificial Neural Networks. Section 3 describes the experimental datasets used for model development and validation. The proposed fatigue life prediction model based on artificial neural networks, along with potential modifications, is introduced in Section 4. Section 5 presents the model's prediction accuracy on the experimental datasets. Finally, the conclusions and discussion are given in Section 6.

## 2 Artificial Neural Networks

The complexity of Artificial Neural Networks (ANNs) arises from the interaction of many simple components known as artificial neurons. Each artificial neuron computes its output $y$ from inputs $\vec{x}$ using the equation:

$$y = g\left(\sum_{i=1}^{N} w_i x_i + b\right), \tag{1}$$

where $\vec{w}$ are the weights, $b$ is the bias and $g(\ldots)$ is the activation function, which introduces the necessary non-linearity into the model. To structure the connections within a network, neurons are organized into substructures called layers. This work employs two types of layers: the Fully Connected (FC) layer and the Convolutional

(CONV) layer. The FC layer consists of a set of neurons where each neuron is connected to every element of the input vector, operating in parallel. The CONV layer can be viewed as applying a smaller FC layer, referred to as a convolutional kernel, to subsections of the input vector. The number of inputs to this kernel is called the filter size. The kernel slides along the input vector in fixed steps, known as the stride, computing a partial output at each position. These partial outputs are then combined to form the output vector of the layer. In this work, the stride is always set equal to the filter size.

# 3 Experimental datasets

The experimental datasets used for model development and validation consist of 20 specimens of 316L stainless steel [1] and 8 specimens of AlSi10Mg aluminium alloy [2], all fabricated using laser powder bed fusion. Prior to fatigue testing, defects within the specimens were characterised using X-ray computed tomography. Based on literature [3], three key defect characteristics were selected. To quantify the size of defects, the Murakami parameter, $\sqrt{area}$, is employed. This is defined as the square root of the projected area on the plane experiencing maximum tensile strain [4]. Defect morphology is described using sphericity, $\varPsi$, which is the ratio of the surface area of a sphere with the same volume to the actual surface area of the defect. Finally, the shortest distance from the defect to the free surface, $L$, is used to capture the positional aspect of the defect [1, 2].

For low-cycle fatigue testing, the 316L specimens were manufactured with a tubular geometry and tested under axial-torsional loading. Of the 20 specimens, 7 were loaded axially, 8 torsionally, and 5 under proportional loading with a strain ratio of $\gamma/\epsilon = \sqrt{3}$. In contrast, the AlSi10Mg specimens were cylindrical, as they underwent purely axial loading. In the evolution of equivalent stress amplitudes over cycles, a tangent line is drawn over the linear portion of the curve. The number of cycles to failure, $N_f$, is then defined as the point where the amplitude falls below 90% of this tangent line. Detailed information on specimen preparation, tomography, and fatigue testing procedures can be found in [1] for the 316L dataset and [2] for the AlSi10Mg dataset.

# 4 Proposed model framework

## 4.1 Conventional model

The development of the proposed framework begins with conventional fatigue lifetime models. Many widely used fatigue models can be generalized by the following equation:

$$D = C_1 N_f^{C_2} + C_3 N_f^{C_4}, \tag{2}$$

where $N_f$ is the fatigue lifetime, $D$ is a damage parameter characterizing the loading cycle, and $C_i$ are material dependent parameters. In conventional models, some of these coefficients may carry physical significance. However, in this work, they are determined numerically through data fitting. The damage parameter $D$ is straightforward to assess under purely axial loading. For multiaxial loading scenarios, additional techniques such as the critical plane method are required to accurately evaluate.

## 4.2 Artificial Neural Network based model

To implement the damaging effect of material defects, the damage parameter is modified by a corrective parameter $\varphi$, leading to the following hybrid form [1, 2]:

$$\varphi D = C_1 N_f^{C_2} + C_3 N_f^{C_4}, \tag{3}$$

where $\varphi$ is to be computed using an ANN based on defect characteristics. This corrective parameter has a lower bound of 1, corresponding to defect-free material, but no fixed upper bound. Other forms of conventional fatigue lifetime models may also be employed, their integration is a matter of how to incorporate the corrective parameter.

The proposed ANN consist of a single CONV layer followed by a stack of FC layers, with the last one being a single neuron [1, 2], as is shown in Fig. 1. The input vector of the ANN includes the logarithm of the Murakami parameter, $ln\left(\sqrt{area}\right)$, sphericity, $\varPsi$, and the distance from free surface, $L$, as defined in Section 3.

The reasoning for applying the logarithmic transformation to the Murakami parameter will be discussed later. To ensure a consistent input size for all data points, defects are sorted in descending order based on their Murakami parameter and only the top $N$ largest defects are used as input [1, 2]. The architecture includes several integer-valued hyperparameters, such as the number of defects $N$, the number of FC layers following the CONV layer K, and the number of neurons in each layer.

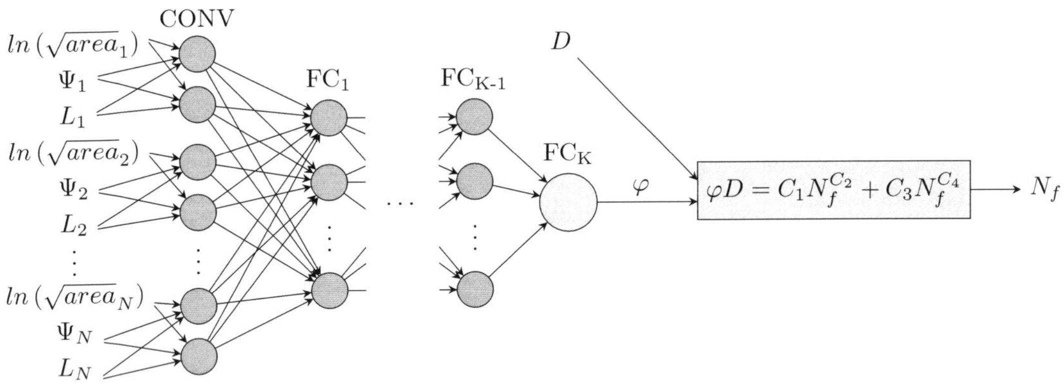


Fig. 1: Diagram of the proposed physics-informed neural network based fatigue lifetime model.

### 4.3 Implementing physical properties

As described so far, the proposed model is capable of predicting fatigue life based on the damage parameter and defect characteristics. However, it presents certain issues. The use of a CONV layer as the first layer instead of FC layer is intended to prevent the ANN from focusing exclusively on a specific member of the input vector while ignoring others. Nevertheless, this issue may re-emerge in the subsequent layer, where the ANN could potentially learn to prioritize specific defects over others. To ensure that each defect is processed in the same manner, the weights in the first layer ($FC_1$) are tied, such that:

$$w_{ij}^{(FC_1)} = w_{ij+kM}^{(FC_1)} \qquad \text{for all } i, j \text{ and } k, \tag{4}$$

where $M$ is the output size of the convolutional kernel in the CONV layer [1, 2]. Further improvements are introduced by enforcing physical constraints on the corrective factor $\varphi$. The first set of constraints stems from the physical intuition that defects are more damaging when they are larger, less spherical, and closer to the free surface:

$$\frac{\partial \varphi}{\partial \sqrt{area_i}} > 0, \quad \frac{\partial \varphi}{\partial \Psi_i} < 0, \quad \frac{\partial \varphi}{\partial L_i} < 0, \tag{5}$$

where $i$ is the index of the defect. The second set of constraints ensures boundary behaviour for $\varphi$, specifically that:

$$\lim_{\forall i(\sqrt{area_i} \to 0)} \varphi = 1, \qquad \lim_{\exists i(\sqrt{area_i} \to \infty)} \varphi = \infty. \tag{6}$$

These conditions guarantee that $\varphi = 1$ for defect-free materials and grows unbounded with increasing defect severity. To satisfy these constraints across all possible inputs, the following emerged [1, 2]:

- The softplus function is selected as the activation function for all layers, except the final layer ($FC_K$), which uses an identity (linear) activation.
- A logarithmic transformation is applied to the $\sqrt{area}$ input.
- In the CONV layer, weights associated with the $ln(\sqrt{area})$ input must be positive and weights associated with $\Psi$ and $L$ inputs must be negative.
- All weights in the FC layers must be positive.

These restrictions ensure compliance with the physical properties defined in Eq. 5 and Eq. 6. To enforce the lower boundary ($\varphi = 1$ for defect-free material), the ANN is evaluated with all $\sqrt{area}$ values set to zero, producing an intermediate output $\varphi_0$. The bias of the final FC layer is then adjusted as:

$$b_k = 1 - \varphi_0, \tag{7}$$

ensuring that $\varphi = 1$ in the absence of defects [1, 2].

### 4.4 Optimization process

The optimization process of the ANN based models is structured in two layers: the inner layer, which optimizes weights $\boldsymbol{w}$, biases $\vec{b}$, and material parameters $C_i$ and the outer layer, which searches for integer-valued hyperparameters. The dataset is typically divided into training, validation, and testing subsets with approximately a 70/15/15% ratio. The training dataset is used in the inner layer to optimize the model parameters. Since the gradient of the model output with respect to the ANN inputs is analytically tractable, stochastic gradient descent methods are well-suited for optimization. In this work, the Adam algorithm is employed, with settings as described in [1], to minimize the following loss function:

$$Loss = \sum_{i=1}^{n} \left(ln\left(y_i\right) - ln\left(t_i\right)\right)^2 + f\left(\boldsymbol{w}\right), \tag{8}$$

where $y_i$ is the fatigue lifetime predicted by the model, $t_i$ is the experimentally measured fatigue lifetime, and $f(\ldots)$ denotes a penalty function. This function enforces the physical constraints on the weights as described in Subsection 4.3, and is zero when all constraints are satisfied [1, 2].

The validation dataset is used in the outer layer to tune the hyperparameters. A grid search is conducted over predefined ranges for each hyperparameter. For every combination, the inner optimization is executed and model performance is evaluated. After completing the grid search, the combination that yields the best performance on the combined training and validation datasets is selected to prevent overfitting. Finally, the testing dataset is used exclusively after all optimization is completed, to evaluate the generalization capability of the model on unseen data.

The optimization process for the conventional models is relatively straightforward. These models are sufficiently robust that splitting the data into training, validation, and testing subsets is not strictly necessary, but the division is kept to ensure consistency in model comparison. Since the conventional models do not involve hyperparameters, the validation subset is merged with the training data. The material parameters $C_i$ are then optimized on the training set using a genetic algorithm with the same loss function (Eq. 8), but without the penalization term.

## 5 Results

The results of the proposed model are presented first for the 316L stainless steel dataset [1]. Based on existing literature, the Smith-Watson-Topper (SWT) parameter [5] is considered the most suitable for this material:

$$D = P_{SWT} = \sigma_{max}\varepsilon_a, \tag{9}$$

where $\sigma_{max}$ and $\varepsilon_a$ are the maximum normal stress and the normal strain amplitude, respectively, acting on a plane experiencing maximum tensile strain, and are evaluated on the mid-life cycle, $N_f/2$. This parameter is part of the critical plane class of fatigue models. The model was optimized according to the procedure described in Subsection 4.4 and the resulting optimal hyperparameters are summarized as follows [1]:

- The prediction accuracy increased with the number of considered defects $N$, until reaching a plateau at $N = 10$, which is therefore the value used in further analysis.
- A stack of two fully connected (FC) layers was found sufficient ($K = 2$).
- The optimal number of neurons was determined to be $M_{CONV} = 2$ and $M_{FC_1} = 1$.

The fatigue lifetime prediction results are shown in Fig. 2. The shown performance metrics Mean Squared Error (MSE) and coefficient of determination ($R^2$) were computed over the whole dataset in the logarithmic scale [1].

The optimization of the proposed model for the AlSi10Mg aluminium alloy dataset posed additional challenges due to the small number of available data points [2]. This dataset used axial strain-controlled loading, which allows for a simpler damage parameter derived from the Coffin-Manson model [6], based on the plastic strain range, $\Delta\varepsilon_{pl}$, which is also evaluated on the mid-life cycle, $N_f/2$:

$$D = \Delta\varepsilon_{pl}. \tag{10}$$

With only 8 specimens, the risk of overfitting was high, as overfitting tends to occur when the number of learnable parameters exceeds the number of training samples. To avoid this, several modifications were introduced

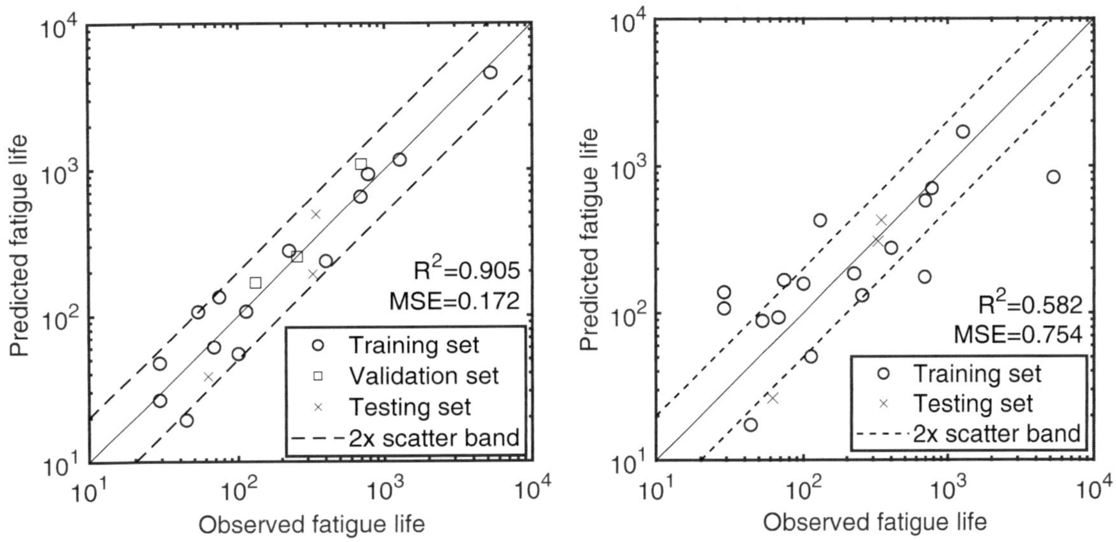


(a) (b)

Fig. 2: Prediction accuracy for the 316L stainless steel dataset of the proposed Physics-Informed Neural Network based fatigue model (a) in comparison to a conventional model (b), which does not consider defects, i.e. $\varphi$ is set to 1 [1].

to reduce the model complexity. The original Coffin-Manson model includes only one power-law term in Eq. 2, therefore $C_3$ is set to zero, leaving only $C_1$ and $C_2$ material parameters. During model development, the network consistently assigned low weights to the sphericity input $\Psi$, hence these features were removed from the input vector. The final ANN architecture was manually reduced to a single neuron in the CONV layer, followed by the output FC layer with one neuron by default [2]. These changes reduced the number of learnable parameters to 6, fewer than the number of data points. With other hyperparameters fixed based on prior results, only the number of considered defects $N$ was optimized, and the optimal value was again found to be $N = 10$. The corresponding lifetime prediction performance is presented in Fig. 3, with the same evaluation metrics as for the 316L dataset [2].

## 6 Conclusion

A Physics-Informed Neural Network based framework was proposed to predict fatigue lifetime model while considering the damaging effect of defects. The model predictive accuracy has been demonstrated using a dataset of 316L stainless steel specimens [1], shown in Fig. 2, and for a smaller dataset of AlSi10Mg aluminium specimens [2], shown in Fig. 3. In both cases, a good correlation between experimental and predicted fatigue lifetimes was observed. Notably, prediction accuracy on the testing sets remained consistent with the rest of the data, suggesting that the models generalize well and do not suffer from overfitting [1, 2].

Future development of the proposed framework could proceed in several directions, such as considering the interference between individual neighbouring defects, which are currently treated independently. More critically, the model is not yet suitable for real components, where the stresses are not constant across the whole component. Promising approaches to address this include statistical analysis on localized sub-regions or incorporating the local damage parameter at each defect directly into the input vector. However, both solutions would significantly increase model complexity and would require a much larger volume of data, which remains a major challenge in this domain.

## Acknowledgement

The authors acknowledge support from the Czech Science Foundation Grant No. 24-11308S, and from the Grant Agency of the Czech Technical University in Prague, No. SGS24/122/OHK2/3T/12.

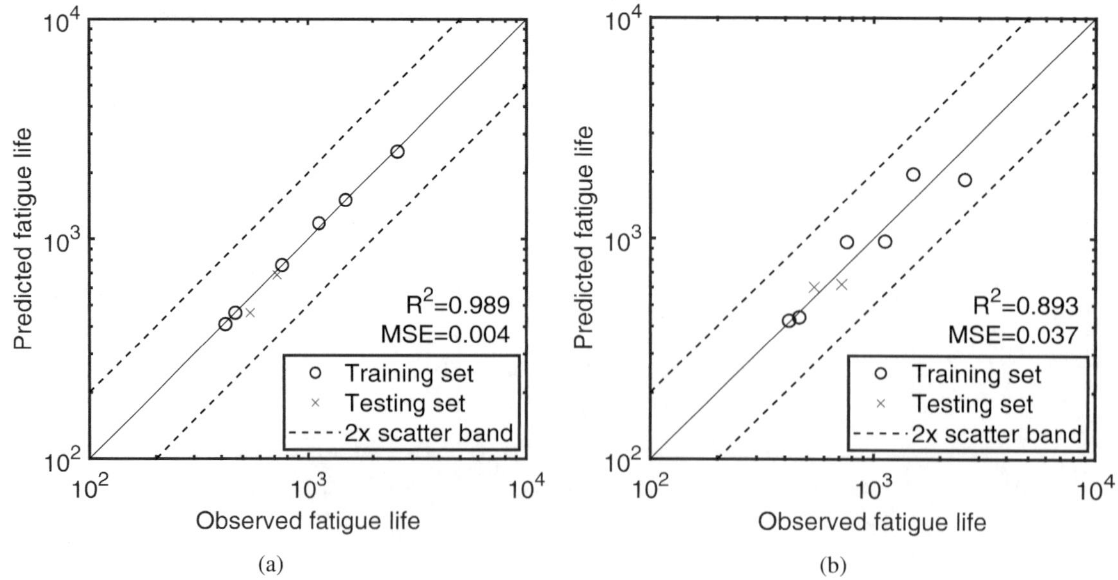


Fig. 3: Prediction accuracy for the AlSi10Mg aluminium dataset of the proposed Physics-Informed Neural Network based fatigue model (a) in comparison to a conventional model (b), which does not consider defects, i.e. $\varphi$ is set to 1 [2].

# References

[1] M. Bartošák et al., Using physics-informed neural networks to predict the lifetime of laser powder bed fusion processed 316L stainless steel under multiaxial low-cycle fatigue loading, International Journal of Fatigue 190 (2025) 108608, doi: 10.1016/j.ijfatigue.2024.108608.

[2] M. Bartošák et al., Low-cycle fatigue of laser powder bed fusion-processed AlSi10Mg using recycled powder: Experiments and machine learning-assisted lifetime prediction, Materials & Design 253 (2025) 113926, doi: 10.1016/j.matdes.2025.113926.

[3] Y. Nadot, Fatigue from Defect: Influence of Size, Type, Position, Morphology and Loading, International Journal of Fatigue 154 (2022) 106531, doi: 10.1016/j.ijfatigue.2021.106531.

[4] Y. Murakami, S. Beretta, Small defects and inhomogeneities in fatigue strength: experiments, models and statistical implications, Extremes 2 (1999) 123–147, doi: 10.1023/A:1009976418553.

[5] K.N. Smith, T.Topper, P.Watson, A stress-strain function for the fatigue of metals, Journal of materials 5 (1970) 767–778.

[6] L.F. Coffin, A study of the effects of cyclic thermal stresses on a ductile metal, Transactions of the American Society of Mechanical engineers 76 (1954) 931–949, doi: 10.1115/1.4015020.

# The Effect of Printing Direction on the Properties of Thermoplastic Elastomers

M. Hančinová [1,a], E. Dubňanská [1], R. Huňady [2], P. Palička [1]

[1] *Technical University of Košice, Faculty of Mechanical Engineering, Slovakia*
[2] *VŠB - Technical University of Ostrava, Faculty of Mechanical Engineering, Czech Republic*
[a] *martina.hancinova@tuke.sk*

**Abstract:** This study investigates how the orientation of printed layers affects the tensile properties of FiberFlex 30D, a flexible thermoplastic elastomer used in 3D printing. Due to the layer-by-layer process of 3D printing, mechanical behaviour can vary depending on how the layers are aligned. In this study, samples were printed in three directions *(x, y, z)* using a Bambu Lab X1C printer based on Fused Filament Fabrication (FFF). Tensile samples were prepared according to ČSN ISO 37 and tested on a universal testing machine at room temperature. The results showed that print orientation significantly influenced the tensile strength and stiffness of the material. The highest strength was observed in samples printed in the y-axis, while those printed in the z-axis showed the weakest performance. These findings confirm the importance of print orientation in the design of flexible 3D-printed parts and provide useful data for improving the structural performance of components made from thermoplastic elastomers.

**Keywords:** thermoplastic elastomer, 3D printing, print orientation, tensile properties, Fused Filament Fabrication (FFF)

## 1 Introduction

3D printing is a modern manufacturing method that enables the production of complex geometric objects from a wide range of materials. In recent years, elastomers – flexible, rubber-like materials – have been increasingly used thanks to their elasticity and durability. However, their mechanical properties largely depend on the printing conditions, particularly the print orientation, which determines the direction of layer deposition and thus the final strength of the object [1].

While many studies have focused on rigid thermoplastics such as PLA or ABS, thermoplastic elastomers (TPEs) remain relatively unexplored [2].

Engineering materials can exhibit anisotropic behaviour, meaning their mechanical properties vary depending on the direction of loading. This anisotropy may arise from intrinsic factors, such as thermal history during crystallization in metals or polymerization in polymers or may be introduced by the manufacturing process itself. In the context of 3D printing, anisotropy is typically process-induced, resulting from the characteristic layer-by-layer deposition of material [2].

The mechanical properties of 3D printed parts, particularly the role of print orientation in determining strength and failure, have been examined in several recent studies.

Adrover-Monserrat et al. (2023) analysed the influence of printing direction on PEBA-polyether-block-amide-based polymer by using tensile testing and Acoustic Emission analysis to identify how failure mechanisms vary with orientation [2].

Similarly, Telu et al. (2024) evaluated the effect of print direction on Filaflex SEBS using a Universal Testing Machine. Their study focused on assessing mechanical parameters such as tensile strength, Young's modulus, and elongation at break to investigate how print orientation influences the behaviour of thermoplastic elastomers [3].

In another study, Omar & Hamid (2024) explored the mechanical behaviour of polymeric models produced via LCD 3D printing. Combining tensile and compression tests with Finite Element Method (FEM) simulations, they found that the On-Edge orientation provided the highest tensile strength, likely due to finer layer resolution and reduced porosity. Conversely, the Up-Right orientation resulted in the lowest strength because the layers were perpendicular to the loading direction. In compression, vertically printed samples demonstrated superior mechanical performance compared to horizontally printed ones [4].

In addition to understanding mechanical properties of material, it is essential to gather data suitable for constitutive modelling. Thermoplastic elastomers often exhibit nonlinear and time-dependent behaviour, making them ideal candidates for hyperelastic and viscoelastic material models. Such models are essential for accurate finite element simulations, which help optimize the design and performance of flexible components. The present study serves as the basis for future model development by providing directional tensile data, and will be followed by compression and relaxation testing to characterize the full mechanical response of the material.

## 2 Experimental Tensile Testing

The tested samples were manufactured using FiberFlex 30D, a thermoplastic elastomer filament developed by Fiberlogy. This material is known for its high flexibility, impact resistance at low temperatures, and durability under mechanical stress. It has a Shore hardness of 30D and is suitable for applications such as gaskets, flexible hinges, tires, rubber components, and wearable items [5]. According to the manufacturer, the optimal nozzle temperature is between 200–220 °C and the bed temperature ranges from 50 to 70 °C.

All samples were printed on a Bambu Lab X1C printer using Fused Filament Fabrication (FFF) technology and a filament diameter of 1.75 mm. The nozzle diameter was 0.4 mm, and the layer height was set to 0.1 mm. The shape and dimensions of samples were determined according to ČSN ISO 37 (62 1436) [6] - Determination of tensile properties, the geometric model of 3D printing samples (Fig. 1).

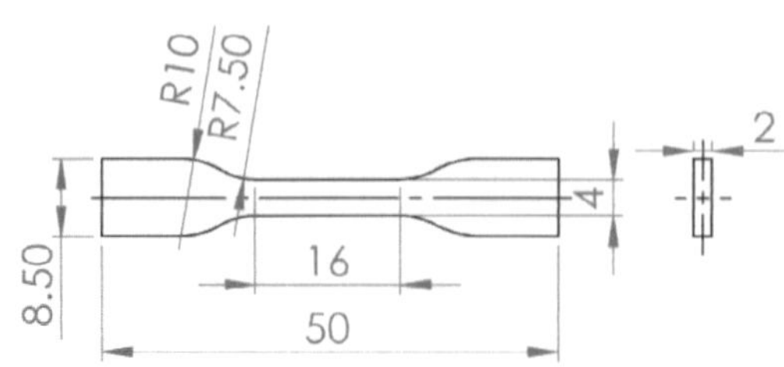


Fig. 1 Dimensions [mm] of 3D printing samples for uniaxial tensile test.

The samples were printed in the $x$, $y$ and $z$ axis direction (Fig. 2) representing different layer alignments with respect to tensile load.

A total of 31 samples were printed: 9 in x, 9 in y, and 13 in z directions. Based on visual inspection and curve analysis, 9 x-axis, 7 y-axis, and 10 z-axis samples were selected for final evaluation. Outliers with premature failure, irregular deformation, or measurement noise were excluded to ensure data consistency.

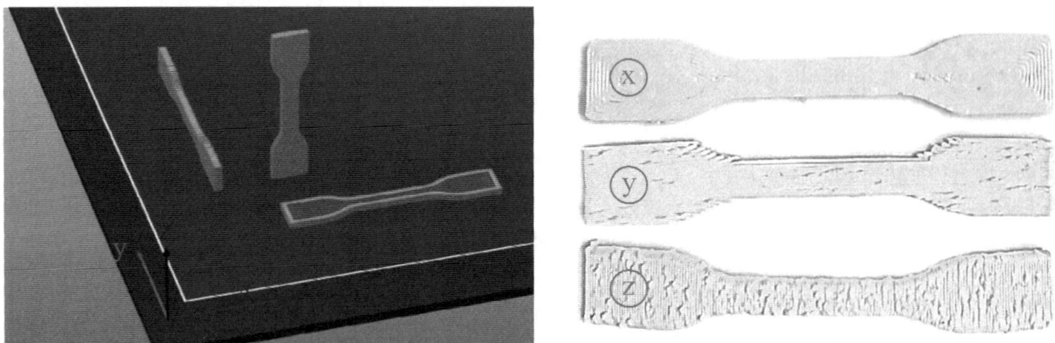


Fig. 2: Schematic of three print orientations and representative samples printed in $x$, $y$, and $z$ directions.

The uniaxial tensile tests were performed on Universal Testing Machine (Kason WDW-5). The loading speed was 200 mm/min. The tests were carried out at constant temperature 23 °C.

The deformation of the samples was measured using an optical video extensometer system (Fig. 3). The method is based on tracking the movement of two reference markers placed on two parallel lines on the surface of the sample, aligned with the tensile axis. During the test, a high-resolution grayscale video was recorded, and the deformation was evaluated using digital image correlation (DIC).

For the analysis, a custom-developed MATLAB interface was used. The program processed the video sequence by automatically correlating small image regions (facets) around each marker in every frame. From this, the time-dependent displacement between the markers was calculated and converted into engineering strain. Combined with the force data from the testing machine and the known cross-sectional area, the corresponding engineering stress was determined.

This contactless measurement approach allows accurate tracking of large deformations without interfering with the sample, making it especially suitable for highly stretchable materials like thermoplastic elastomers.

Fig. 3 shows the optical measurement setup and evaluation results. The left image displays the camera monitoring the sample during testing. The right side presents screenshots from the analysis software: the original sample image, the placement of tracking markers, and the stretched sample with two tracked points highlighted.

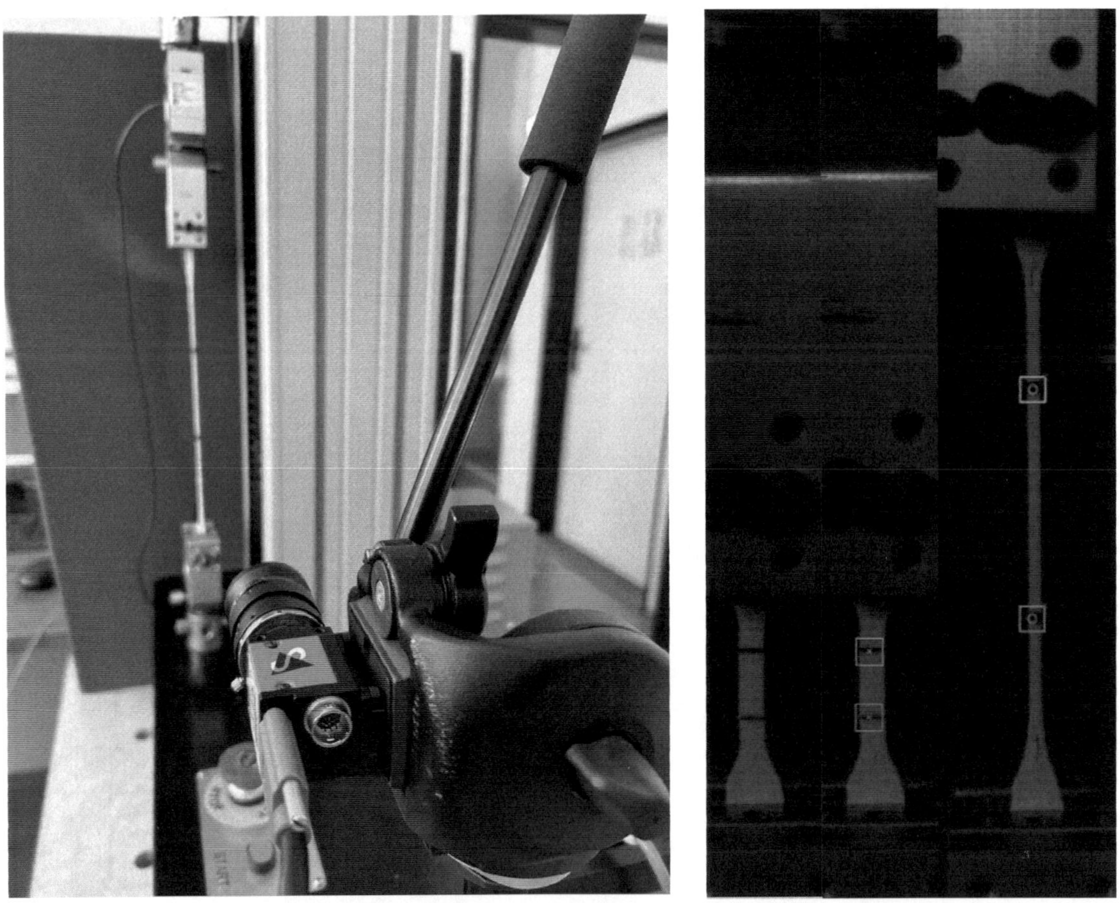

Fig. 3 Measurement of sample deformation using a video extensometer.

The stress–strain curves shown in Fig. 4 represent averaged data from the evaluated samples in each orientation. The data clearly show distinct mechanical behaviour depending on print orientation, which highlights the anisotropy introduced by the layer-wise deposition. To quantify these differences, the Young's modulus $E$ was determined from the initial linear region of the curves, corresponding to small strains up to 5 %. The slope of this region represents the small-strain elastic response, and the quality of the linear fit was evaluated using the coefficient of determination $R^2$, which reached a value of 0.99 for all three printing orientations. The resulting values of $E$ for each printing orientation are summarized in Tab. 1.

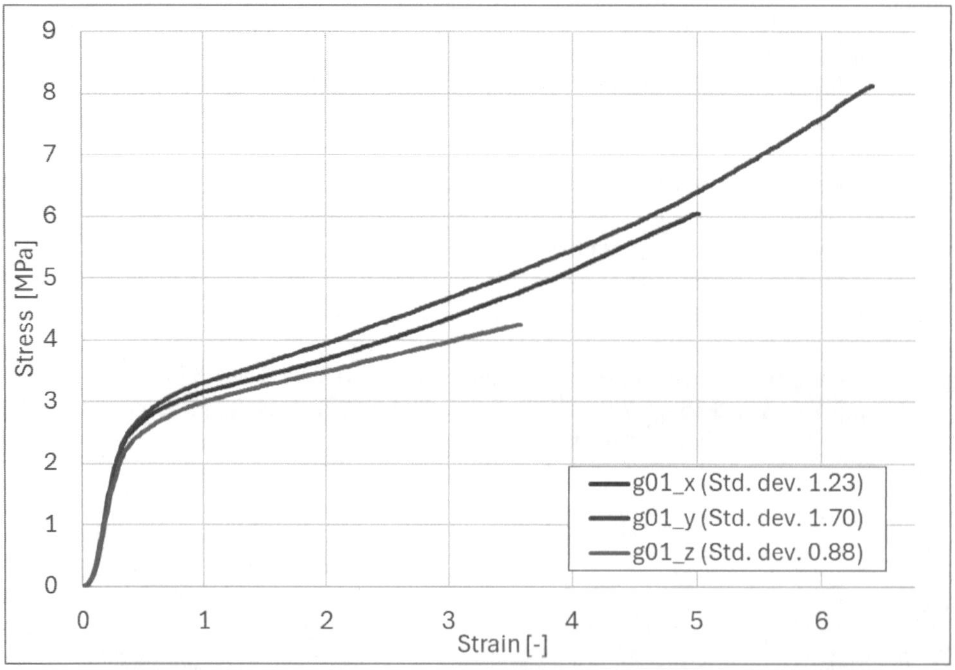


Fig. 4 Comparison of stress-strain curves of samples printed in $x$, $y$ and $z$ directions. Values in the legend indicate the corresponding standard deviations of the measured data.

Tab. 1 Young's modulus of printed samples in different orientations.

| sample | **g01_x** | **g01_y** | **g01_z** |
|---|---|---|---|
| $E$ [MPa] | 9.39 | 9.67 | 8.64 |

In addition to the numerical results, some noticeable differences were observed in how the samples behaved during testing, depending on their printing orientation. Samples printed in the $z$-direction often failed earlier and along the layer boundaries, suggesting weaker bonding between layers. These failures were typically abrupt and occurred along interlayer boundaries, indicating brittle behaviour. On the other hand, $y$-direction samples stretched more gradually and failed in a smoother, more uniform way, indicating stronger interlayer adhesion when the layers are aligned with the load.

These observations show that printing direction doesn't just affect the strength or stiffness of the material, but also how it deforms and breaks under load. In some cases, the samples failed suddenly and without visible necking, others exhibited significant elongation and stable deformation before breaking.

## 3 Conclusion

The tensile test results confirm that layer orientation significantly influences the mechanical properties of FiberFlex 30D material. The highest strength was observed in the samples printed in the $y$-axis direction, indicating an optimal bond between the layers in this orientation. Samples printed in the $x$-axis and $z$-axis exhibited lower stiffness and strength, with the z orientation being the weakest. This behaviour is consistent with expectations that samples with layers oriented perpendicular to the loading direction will fail first. In the $y$-axis orientation, strength is dominated by the continuous extruded filament being aligned with the load, whereas z-axis strength is limited by the weaker interlayer adhesion.

These findings are in line with previous research on other 3D-printed polymers and emphasize the role of process-induced anisotropy in elastomeric materials. Understanding how print orientation affects mechanical behaviour is especially important in the design of flexible parts where directional strength plays a key role, such as in seals, damping elements, etc.

As a next step, the research will be extended to include compression and relaxation tests, which will provide a more comprehensive understanding of the material's behaviour under different types of mechanical loading. Compression testing will help characterize the response of the material under confined stress states, which are critical in applications involving contact or deformation under pressure. Stress-relaxation experiments will provide time-dependent mechanical data necessary to describe viscoelastic behaviour, which is typical for elastomeric materials.

The data obtained from these tests will serve as the basis for developing a constitutive material model capturing the nonlinear, anisotropic, and time-dependent behaviour of FiberFlex 30D. This model will be implemented in FEA software to support the simulation and optimization of flexible 3D-printed components.

## Acknowledgement

This work was supported by the Slovak Research and Development Agency under the grant project VEGA No. 1/0516/22, and by APVV-23-0461, and by European Union under the REFRESH–Research Excellence for REgion Sustainability and High-tech Industries project number CZ.10.03.01/00/22_003/0000048 via the Operational Programme Just Transition and by the EU NextGenerationEU through the Recovery and Resilience Plan for Slovakia under the project No. 09I03-03-V05-00015.

## References

[1] M. Nashruffi, et al. The Effect of Printing Orientation on the Mechanical Properties of FDM 3D Printed Parts. Enabling Industry 4.0 through Advances in Manufacturing and Materials, Lecture Notes in Mechanical Engineering (2022) 75-85, http://doi.org/10.1007/978-981-19-2890-1_8.

[2] B. Adrover-Monserrat, et al. Impact of printing orientation on inter and intra-layer bonds in 3D printed thermoplastic elastomers: A study using acoustic emission and tensile tests. Polymer 283 (2023), https://doi.org/10.1016/j.polymer.2023.126241.

[3] S. B. Telu, A. K. Myadam, and L. R. Kamireddy, Impact on Mechanical Properties of Thermoplastic Elastomers (TPE) Using Different Printing Directions in Additive Manufacturing, SSRG International Journal of Polymer and Textile Engineering, vol. 11, no. 1, pp. 1–5, Mar. 2024. [Online]. Available: https://www.internationaljournalssrg.org/IJPTE/2024/Volume11-Issue1/IJPTE-V11I1P101.pdf

[4] M. Omar and A. Hamid, *Influence of 3D Printing Orientations on Mechanical Properties for Polymeric Models*, ResearchGate, Jan. 2024. [Online]. Available: https://www.researchgate.net/publication/387281195

[5] Fiberlogy, Flexible filament FIBERFLEX 30D – Product Specification, Fiberlogy, https://fiberlogy.com/en/fiberlogy-filaments/flexible-filament-fiberflex-30d/.

[6] ČSN ISO 37 (62 1436) - Rubber, vulcanised or thermoplastic elastomer

# Tensile Properties of 3D Printed Infill Structures with Different Densities

J. Heczko[1,*], J. Krystek[1], V. Laš[1]

[1] *NTIS - New Technologies for the Information Society, Faculty of Applied Sciences, University of West Bohemia, Technicka 8, Pilsen, Czech Republic*

* *jheczko@ntis.zcu.cz*

**Abstract:** The tensile properties of 3D printed rectilinear infill pattern are investigated. Stress-strain curves are obtained for specimens with different infill densities and orientations. The values Young's moduli and ultimate tensile strength are evaluated and polynomial approximations for the Young's moduli are calibrated.

**Keywords:** 3D print; infill density; rectilinear pattern; tensile properties; stress-strain curve

## 1 Introduction

This work is part of a larger effort in modelling, optimization, and design of 3D printed structures. For the current development phase, FDM/FFF printed plastic is used as a cheap way of testing the general ideas and algorithms. Nevertheless, the resulting algorithms and methods are applicable to other industrial grade additive manufacturing systems as well as the extrusion based ones. The present article deals with exploration of the space of material properties. The measured data will serve as the basis of our follow-up work, namely development of methods for structural optimisation of infill structures with the aim of designing and printing an optimal non-uniform sandwich structure. As the most important parameters of the infill structures were identified density and orientation at this stage. The literature already provides similar studies, but often with different materials, infill geometries, or density ranges. To mention a few, Afshar, Jeanne, and Abali [1] tested polylactic acid (PLA) specimens of different infill densities using the ISO 527-2 specimen geometry. Lubombo and Huneault [4] compared five different infill pattern, finding out that square infill has best stiffness and strength in tension, while hexagonal infill has best stiffness and strength in bending. They also mention that stretch-dominated behavior is desired for structural, load-bearing, applications, while bending-dominated structures might be advantageous in energy absorbing applications due to long plateau in the stress-strain relationship.

## 2 Specimen manufacture

The tensile specimens were printed on a Prusa i3 MK3 printer with a 0.4 mm diameter nozzle using a white PETG filament of 1.75 mm diameter made by Prusament. The most important parameters of the printing process are listed in table 1. The shape of the specimens was type I by the standard ISO 1798 [2], see Fig. 1a. Different values of *infill density* were achieved by varying the values of *extrusion width* and *extrusion path spacing*. In addition, the infill patterns were generated under three different angles in order to study their anisotropy.

Fig. 1b shows the parameter space, constraints, and preferred designs. The bounds and constraints for the design variables were based on several considerations:

- *Infill density* should not be larger than 100 %.
- Lower and upper bounds of *extrusion width* are largely determined by nozzle diameter. These constraints have some built-in default values in the slicing software, e.g. for the 0.4 mm nozzle, the *extrusion width* must lie between 0.2 mm and 1.3 mm. However, it was found that the minimum *extrusion width* for the combination of the chosen infill structures, material, and other settings is 0.3 mm. Moreover, the printing

Tab. 1: Printing parameters

| | |
|---|---|
| Layer height | 0.2 mm |
| Nozzle temperature | 250 °C (1st layer 240 °C) |
| Bed temperature | 90 °C (1st layer 85 °C) |
| Max. volumetric speed | 8 mm$^3$/s |
| 1st layer speed | 20 mm/s |

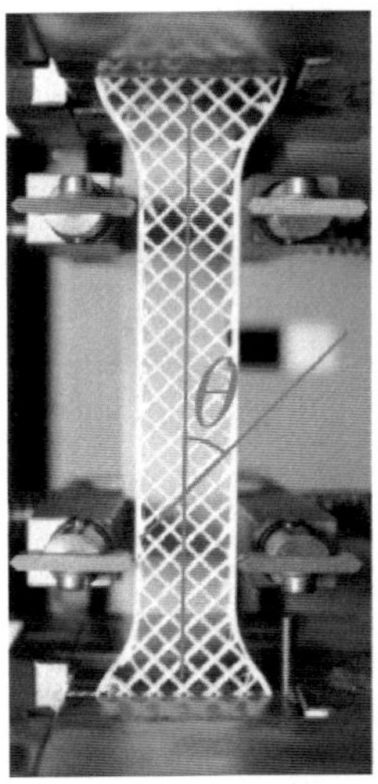

(a) Photo of the specimen, infill angle $\theta$ highlighted

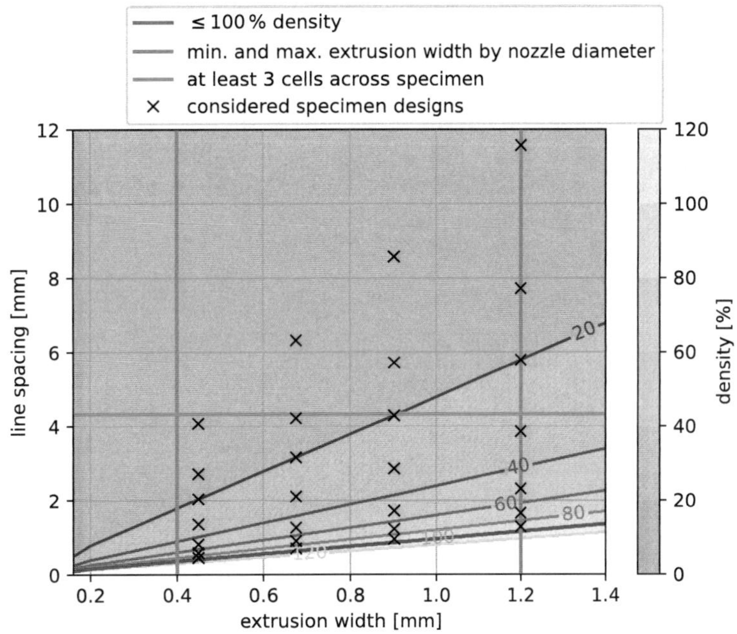


(b) Parameter space with the resulting density values. Various design constraints are highlighted.

Fig. 1: Specimen design.

speed had to be severely reduced in the case of low density specimens (10 % and 15 %). Nevertheless, the printing of tensile specimens with 0.3 mm *extrusion width* was very unreliable, therefore the final lower limit was further increased to 0.45 mm, which is the default value in the slicing software.

- The upper bound of *extrusion path spacing* should result in sufficient number of basic cells across specimen width.

Some of the designs (the lowest density at extrusion widths above 0.45 mm) clearly lie outside of the admissible region – these were omitted from the experiments.

Since it was necessary to use at least a single perimeter for successfuly printing the specimens with accurate geometry as it provides anchor points for the infill extrusion paths that rapidly change direction at the object's boundary. An additional set of testing specimens was therefore manufactured in order to examine the influence of the perimeter. The specimen is shown in Fig. 2. Its infill in the central region is designed so that it adds minimum material in the axial direction, while reinforcing it in the perpendicular direction. The dense infill between the clamps of the testing machine ensures that the specimen is not crushed.

# 3 Experiments

The measurements were carried out on a Zwick/Roell Z050 universal testing machine. Monotonous displacement-controlled loading was prescribed with the speed of 2 mm/min at the clamps. A mechanical extensometer was attached to the central part of the specimen with the initial distance between the extensometer's arms $L_0 = 50\,\mathrm{mm}$.

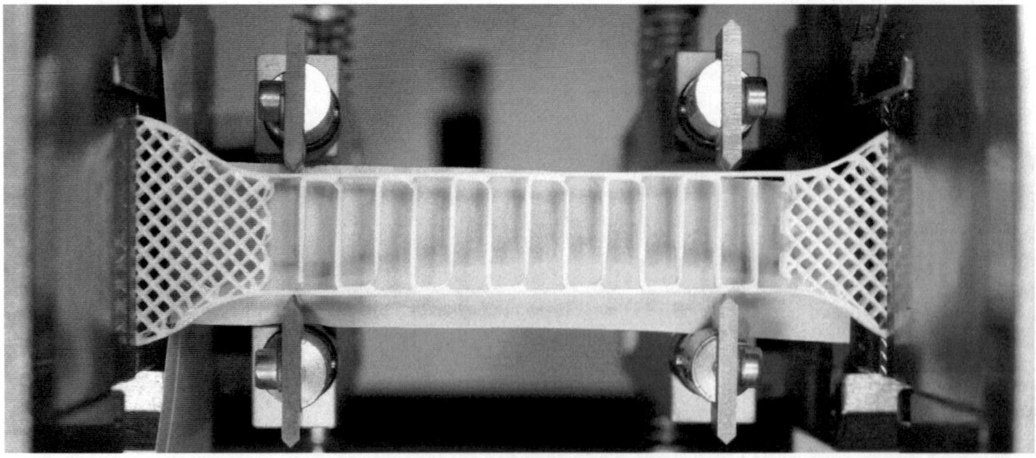

Fig. 2: Single-perimeter specimen.

Tab. 2: Mechanical properties of the material at 100 % infill density as measured by the manufacturer [3].

| | |
|---|---|
| Young's modulus | $(1.5 \pm 0.1)\,\mathrm{GPa}$ |
| Ultimate tensile strength | $(47 \pm 1)\,\mathrm{MPa}$ |

Nominal stress, $\sigma$, was calculated as

$$\sigma = \frac{F}{A_0} \,, \tag{1}$$

where $F$ is the measured force and $A_0 = 4\,\mathrm{mm} \times 13\,\mathrm{mm}$ is the original area of the specimen's outer cross-section. Strain, $\varepsilon$, was evaluated as

$$\varepsilon = \frac{\Delta L}{L_0} \,, \tag{2}$$

where $\Delta L$ is the change in length of the measured part of the specimen and $L_0 = 50\,\mathrm{mm}$ is the original length of the measured part of the specimen.

The *Young's modulus*, $E$, was evaluated as

$$E = \frac{\sigma_2 - \sigma_1}{\varepsilon_2 - \varepsilon_1} \,, \tag{3}$$

where the strain values were fixed as $\varepsilon_1 = 2 \cdot 10^{-4}$, $\varepsilon_2 = 3 \cdot 10^{-3}$ and the corresponding stress values, $\sigma_1$ and $\sigma_2$, were interpolated linearly.

The *ultimate tensile strength* was evaluated as the maximum stress within the data before the first significant fracture. The instant of this first fracture was determined manually by identifying a significant decrease in the measured force.

Table 2 lists mechanical properties as given by the manufacturer's technical datasheet [3], which might be seen as ultimate values. Their measurements were done according to ISO 527-1 and printing the specimens was done at 200 mm/s speed, while other parameters and conditions were very close to those in this article. The values of *Young's modulus* are almost identical to our results, while *ultimate tensile stress* values are significantly higher in [3] than in this article.

## 3.1 Results

The measured stress-strain curves are shown in Fig. 3. The most important observation from these might be that infills oriented at 0° behave in the most brittle way, while increasing the angle increases ductility. The inverse holds for extrusion width, i.e. the lowest extrusion width seems to result in most ductile behavior.

Fig. 4-7 show the results in terms of *Young's moduli* and *ultimate tensile strength*. The comparison specimens with a single perimeter (see the end of Sec. 2) resulted in the average values of *Young's modulus* and *ultimate tensile strength*

$$E = 142.7\,\mathrm{MPa} \,, \quad \sigma_{\mathrm{UTS}} = 2.631\,\mathrm{MPa} \,. \tag{4}$$

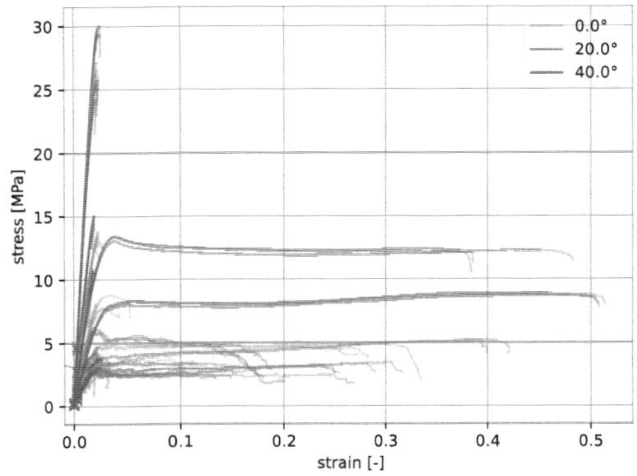


(a) Extrusion width 0.45 mm

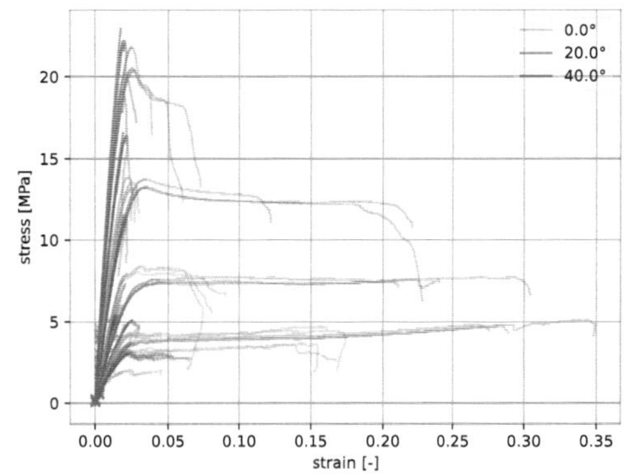


(b) Extrusion width 0.675 mm

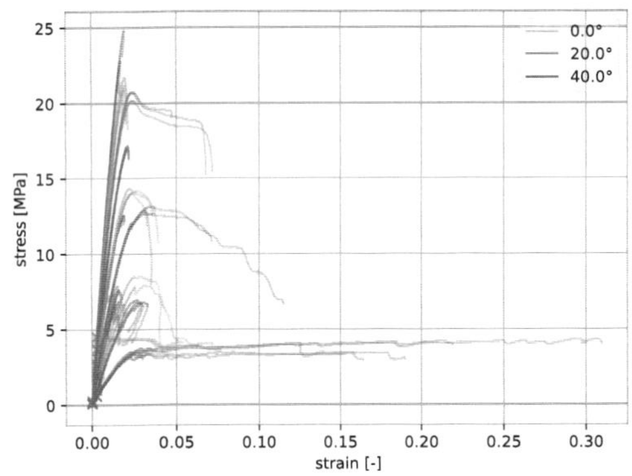


(c) Extrusion width 0.90 mm

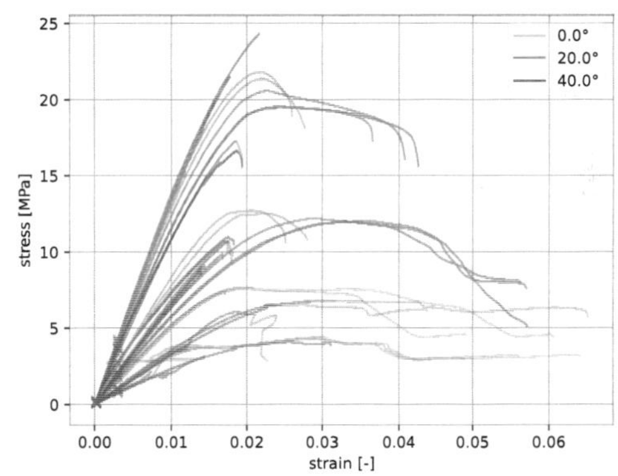


(d) Extrusion width 1.20 mm

Fig. 3: All measured stress-strain curves.

These values align well with the data obtained from the other experiments in the sense that the density-dependent data seem to approach the values in (4) for zero density, with the following exceptions:

- Some specimens with 0° infill orientation. The cause is that the slicing software automatically adds extrusion lines parallel to the perimeters for some low-density specimens. Therefore, some specimens ended up with higher number of fibers in the axial direction than they should according to infill density – see Fig. 8.
- Specimens with high *extrusion width* and low infill density and 20° or 40° infill orientation. Here, the inconsistencies are most probably caused by improper connection between the infill and perimeter in the transition region between the narrow central part of the specimen and the wider end part, see Fig. 9.

### 3.2 Stiffness approximation

In order to use the results for structural optimisation, the stiffness of different infill options was approximated by a continuous function. The measured values of Young's moduli, $E$, were approximated as functions of infill density, $\rho$, by second degree polynomials in the form

$$E(\rho) = E_0 + E_1\,\rho + E_2\,\rho^2\,. \tag{5}$$

Each set of measurements of the same *extrusion width* and *infill angle* were calibrated by a separate set of coefficients $E_0$, $E_1$, $E_2$. The values of the parameters are listed in Table 3 and the comparison to the data is shown in Fig. 10-12. It is expected that $E_0$ would be the same for every structure, since it corresponds to $\rho = 0$, which is equivalent to a specimen with a single perimeter and no infill. Similarly, non-decreasing functions were expected (i.e. $E_1$ and $E_2$ should be non-negative). In several cases, one or the other property is not fulfilled, mostly because of lack of data points at low infill densities.

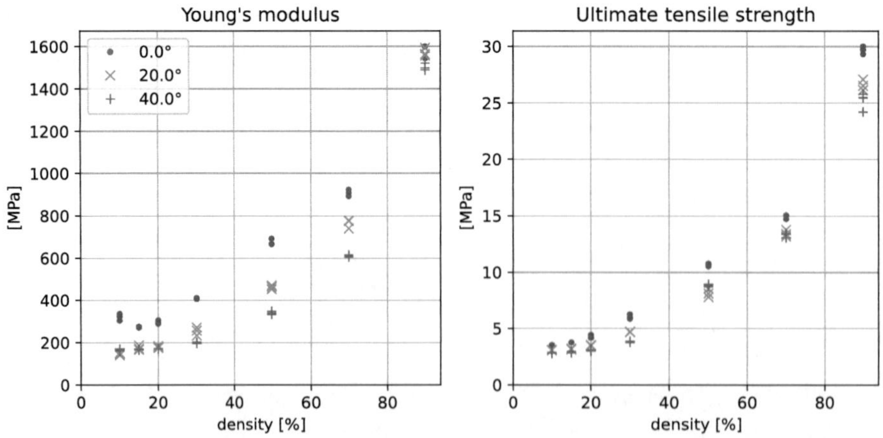


Fig. 4: Stiffness and tensile strength; extrusion line width 0.45mm.

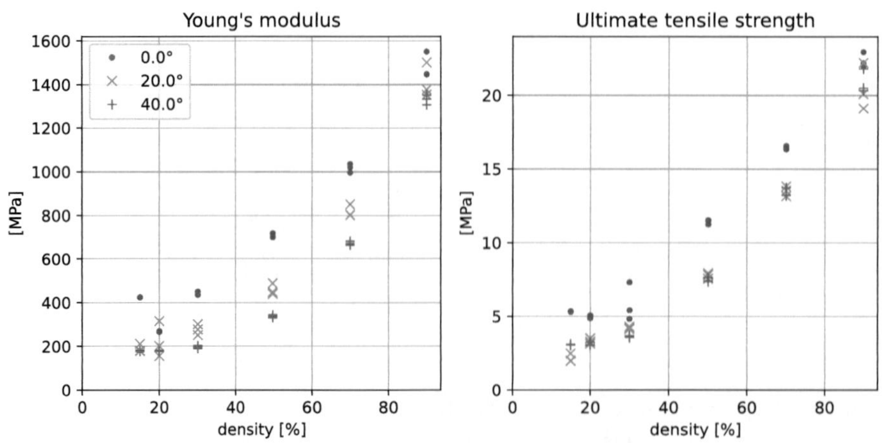


Fig. 5: Stiffness and tensile strength; extrusion line width 0.675mm.

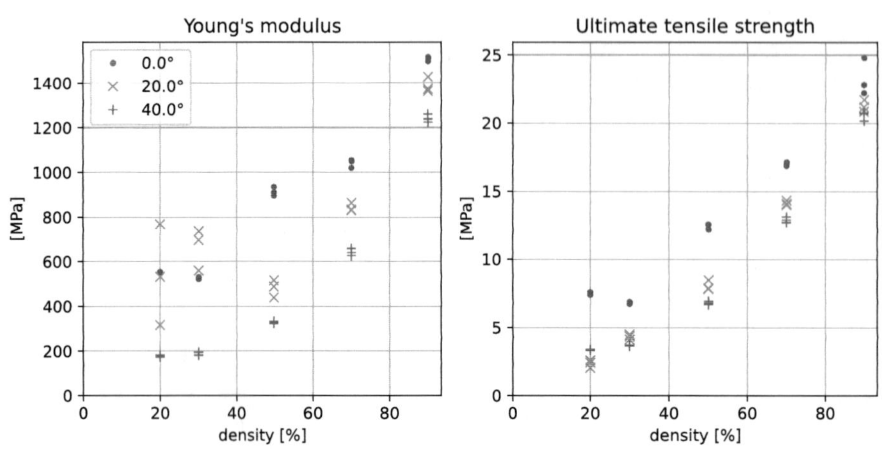


Fig. 6: Stiffness and tensile strength; extrusion line width 0.90mm.

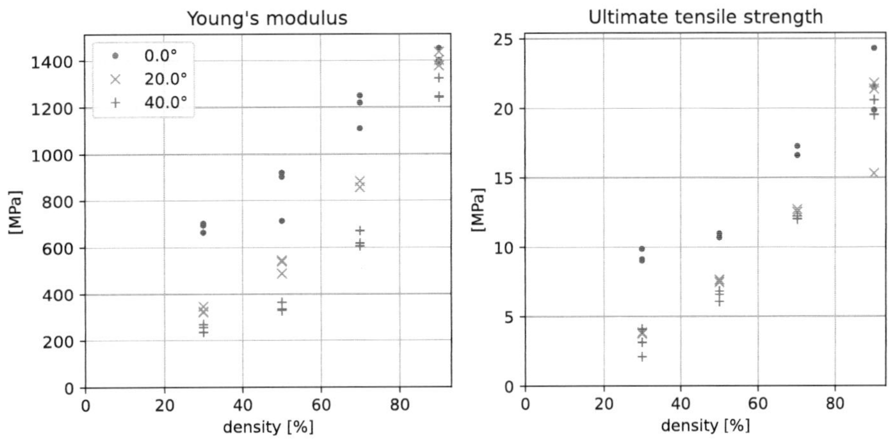


Fig. 7: Stiffness and tensile strength; extrusion line width 1.20mm.

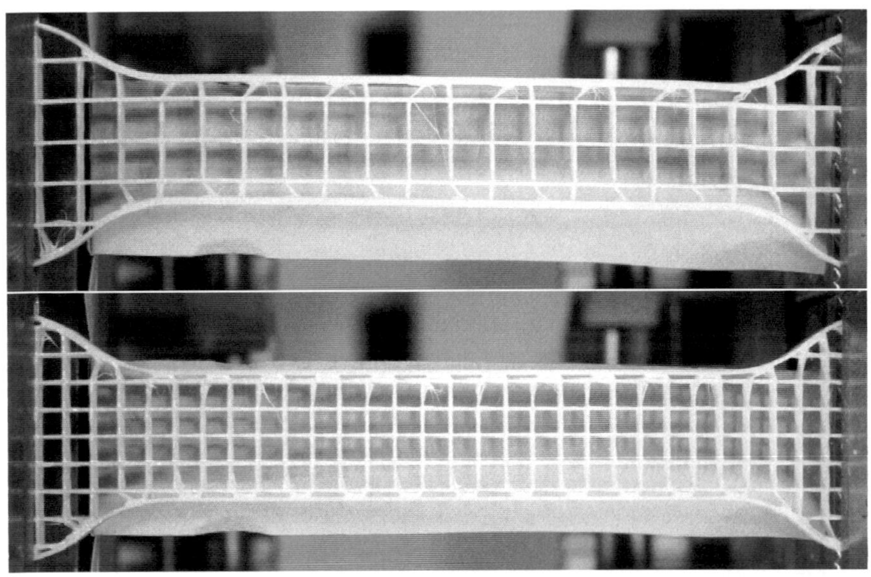

Fig. 8: Low density specimens with extra axial fibers. Top: 10 % infill density, an additional extrusion line is attached to the perimeter along the whole central part of the specimen. Bottom: 15 %. the same total number of extrusion lines along the axis of the specimen.

Fig. 9: Poor bonding between infill and perimeters in case of high extrusion width.

Tab. 3: Parameters of the approximations of Young's moduli (in [MPa]).

| | | infill angle | | | | | | | | |
|---|---|---|---|---|---|---|---|---|---|---|
| | | 0° | | | 20° | | | 40° | | |
| | | $E_0$ | $E_1$ | $E_2$ | $E_0$ | $E_1$ | $E_2$ | $E_0$ | $E_1$ | $E_2$ |
| extrusion width | 0.450 mm | 291.8 | -2.797 | 0.18550 | 262.2 | -10.08 | 0.2678 | 364.0 | -17.70 | 0.3298 |
| | 0.675 mm | 106.1 | 7.677 | 0.08276 | 406.8 | -11.93 | 0.2561 | 355.4 | -14.62 | 0.2809 |
| | 0.900 mm | 424.7 | 2.873 | 0.09937 | 378.2 | -9.422 | 0.2295 | 373.4 | -14.64 | 0.2687 |
| | 1.200 mm | 396.0 | 7.995 | 0.03868 | 369.5 | -7.627 | 0.2123 | 691.5 | -24.68 | 0.3448 |

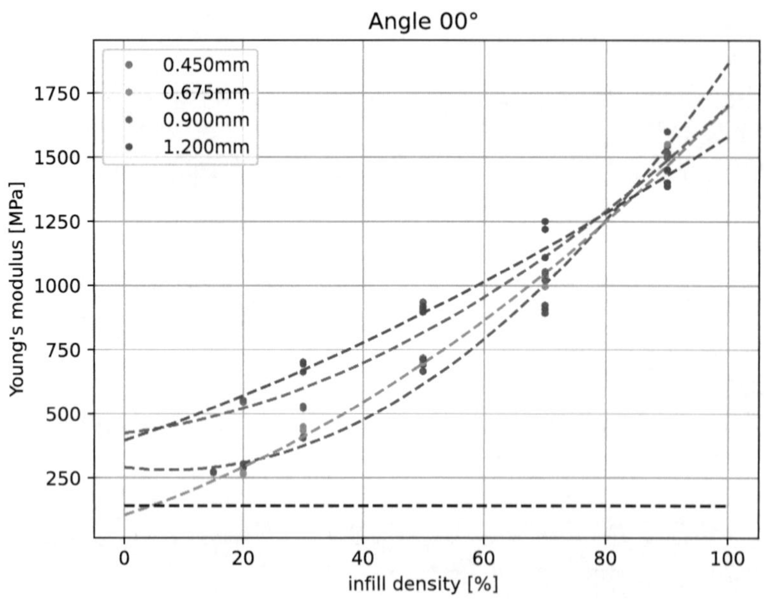


Fig. 10: Stiffness approximation, angle 0°.

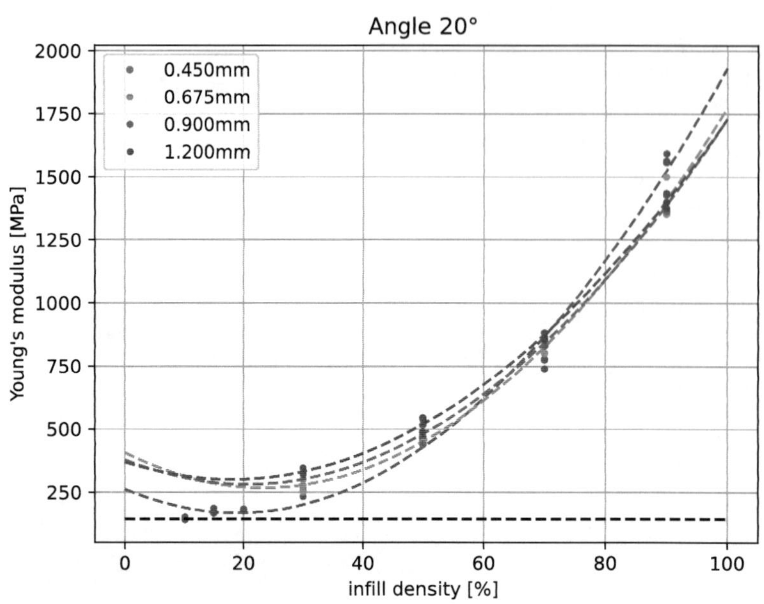


Fig. 11: Stiffness approximation, angle 20°.

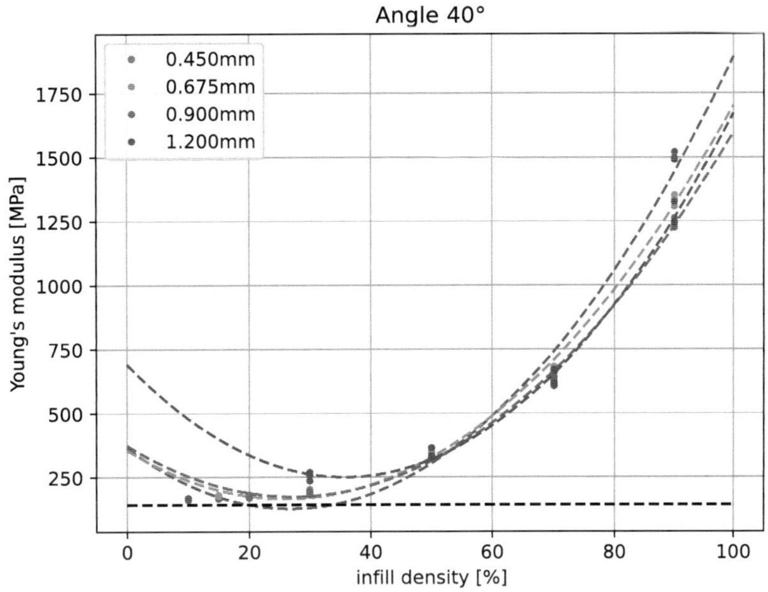


Fig. 12: Stiffness approximation, angle 40°.

# 4 Conclusion

A comprehensive set of experiments was performed in order to study both the manufacturability of the infill patterns at various densities and their stress-strain properties. An admissible range of extrusion widths was verified experimentally. The obtained values of Young's moduli and ultimate tensile strength align well with expectation and data from literature.

Some deficiencies of the present approach were identified, most importantly the shape and size of the specimens. A larger rectangular specimen would enable testing lower infill densities even for high values of extrusion widths. Also, a slicing software with more precise control of the resulting gcode, e.g.[5] would help mitigate some unwanted features of the specimens.

# References

[1] R. Afshar, S. Jeanne, B. E. Abali, Nonlinear Material Modeling for Mechanical Characterization of 3-D Printed PLA Polymer With Different Infill Densities, Applied Composite Materials 30 (2023) 987–1001, doi: 10.1007/s10443-023-10122-y.

[2] ISO 1798 (2008) Flexible cellular polymeric materials – Determination of tensile strength and elongation at break.

[3] Prusa Polymers, Technical datasheet Prusament PETG. Retrieved from: `https://www.prusa3d.com/file/1566978/prusament-petg-technical-data-sheet.pdf`

[4] Ch. Lubombo, M. A. Huneault, Effect of infill patterns on the mechanical performance of lightweight 3D-printed cellular PLA parts, Materials Today Communications 17 (2018) 214–228, doi: 10.1016/j.mtcomm.2018.09.017.

[5] A. Gleadall, FullControl GCode Designer: Open-source software for unconstrained design in additive manufacturing, Additive Manufacturing 46 (2021), doi: 10.1016/j.addma.2021.102109.

# Design of Magnetoactive Soft Gripper Using a Genetic Algorithm

M. Hermann[1,a], I. Petríková[1]

[1] *Technical University of Liberec, Department of Applied Mechanics, Studentská 1402/2, 461 17 Liberec 1, Czech Republic*

[a] *martin.hermann@tul.cz*

**Abstract:** This work presents an approach for designing magnetoactive soft grippers using a simplified magnetomechanical model combined with a genetic algorithm. The studied gripper is a beam-like structure composed of two connected segments made of hard-magnetic elastomer. The proposed method enables fast design of such soft grippers, including their magnetization patterns and magnetic actuation, to achieve desired deformations under static magnetic fields. Compared to full multiphysics simulations, the framework significantly reduces computational time while maintaining good accuracy for predicting the deformation of the soft gripper, making it suitable for rapid prototyping and design exploration of soft magnetic robots.

**Keywords:** soft robots; soft gripper; magnetoactive elastomer; genetic algorithm.

## 1 Introduction

Magnetoactive elastomers (MAEs) are smart composite materials capable of deforming in response to external magnetic fields. In this work, we focus on MAEs containing hard magnetic particles (h-MAEs), which retain a remanent magnetization after exposure to a strong field (typically around 0.8–1 T). This property allows the actuator to deform in a programmable way under external magnetic fields.

The coupled magneto-mechanical behavior of MAEs is strongly nonlinear, which poses significant challenges for simulation and optimization. Standard finite element simulations based on full multiphysics models are often computationally expensive and impractical for iterative design tasks or real-time applications.

To address this, we employ a simplified magnetomechanical model developed in our previous work [1], based on the force acting on magnetic dipole moments in the external field [2]. This reduced model drastically lowers computational cost while capturing the essential behavior of the system.

In the present study, this model is integrated with a genetic algorithm to perform automated design of soft grippers made of h-MAE. The algorithm tunes both the magnetization profile and the actuation field configuration to achieve target deformations under the magnetic field from a permanent magnet.

## 2 Methodology

The overall framework combines a fast magnetomechanical simulation model [1] with a genetic algorithm to explore the design space of magnetically actuated soft grippers. The design parameters include both the magnetization distribution within the h-MAE and the configuration of the external magnetic field.

The method is fully implemented in MATLAB and builds upon a previously developed simplified magnetomechanical model. In this model, each finite element is considered as a magnetic dipole moment, subjected to both magnetic force $\boldsymbol{f}_m$ and torque $\boldsymbol{\tau}$ derived from:

$$\boldsymbol{f}_m = (\boldsymbol{m} \cdot \boldsymbol{\nabla})\boldsymbol{B}_\mathrm{A}, \qquad \boldsymbol{\tau} = \boldsymbol{m} \times \boldsymbol{B}_\mathrm{A}. \tag{1}$$

Here, $\boldsymbol{m}$ is the magnetic dipole moment vector and $\boldsymbol{B}_\mathrm{A}$ is the actuation magnetic field. The direction and magnitude of $\boldsymbol{m}$ are determined experimentally from magnetic hysteresis loop measurements of the h-MAE samples, while $\boldsymbol{B}_\mathrm{A}$ is computed using analytical expressions based on the magnetization, geometry and position of the permanent magnet [3].

## 2.1 Computation of the magnetic dipole moment vector

In the simplified model, each finite element is represented as a magnetic dipole characterized by a magnetic dipole moment vector $\boldsymbol{m}$. The magnitude of the magnetic dipole moment is computed as:

$$\|\boldsymbol{m}\| = M \cdot V_e, \tag{2}$$

where $M$ is the magnetization of the material and $V_e$ is the volume of the finite element, which is known from the mesh generation process.

The magnetization $M$ is determined experimentally based on the relation between the applied magnetic field and the resulting remanent magnetization. This relationship is obtained from magnetic hysteresis loop measurements of the h-MAE material. For each element, the direction of $\boldsymbol{m}$ corresponds to the local magnetization direction assigned as part of the design parameters.

## 2.2 Computation of the actuation magnetic field

The deformation of the h-MAE is controlled by an external magnetic field $\boldsymbol{B}_{\mathrm{A}}$ generated by a permanent magnet. In this work, a magnet with azimuthal symmetry is used, which allows the actuation field to be computed efficiently using analytical expressions [3].

The magnetic field is evaluated as a vector field based on the geometry, position, and magnetization of the magnet. As a simplification, the actuation field $\boldsymbol{B}_{\mathrm{A}}$ is computed solely as the magnetic field generated by the permanent magnet. In reality, $\boldsymbol{B}_{\mathrm{A}}$ is influenced by the presence of the magnetized h-MAE, which itself produces a magnetic field. However, the field induced by the h-MAE is significantly weaker compared to the field from the permanent magnet and is therefore neglected.

## 2.3 Implementation of the genetic algorithm

Genetic algorithms (GAs) are stochastic optimization methods inspired by the principles of natural selection and evolutionary biology. The specific implementation can be modified according to the designer's creativity. The procedure described below follows a workflow inspired by the methodology presented in [4].

Each individual in the population is characterized by three parameters: the magnetization directions $\alpha_1$ and $\alpha_2$, and the horizontal position $x_{\mathrm{mag}}$ of a permanent magnet. The initial population consists of 8 randomly generated individuals. In each generation, a new set of 20 individuals is created using selection, crossover, and mutation.

The selection mechanism is based on fitness-proportional probability: better individuals are more likely to be chosen as parents for the next generation (selection). In addition, the best individual from the current generation is always retained (elitism) and directly passed to the next generation to ensure that the best solution is not lost due to randomness.

Crossover is applied to pairs of selected parents, with the offspring inheriting parameter combinations that are likely to improve the fitness. Mutation is performed by element-wise multiplication of a matrix containing all individuals of the current generation with a randomly generated mutation matrix of the same size. At random positions, the mutation factor is set to 1 (i.e., no mutation), while at other positions, the values are randomly generated within a specified mutation range. For example, if the maximum allowed mutation is $30\%$, the mutation factors vary between 0.7 and 1.3.

The fitness function is evaluated by simulating the resulting deformation and calculating the distance between the tip of the gripper and a predefined target point. The optimization converges within a few minutes.

## 2.4 The workflow of the algorithm

The workflow of the optimization algorithm consists of several key steps, described as follows:

1. **Mesh generation.** A parametric finite element mesh of the soft gripper is generated. This mesh serves as the basis for evaluating the deformation resulting from different actuation parameters.

2. **Optimization setup.** The user specifies which parameters are to be optimized (e.g., magnetization directions, position of the external magnet). The initial population size is defined, along with the number

of individuals to be generated through crossover, the mutation range, and selection strategy. Elitism is enabled to ensure the best solution is preserved across generations.

3. **Evaluation loop.** For each generation, all individuals (candidate solutions) are processed as follows:
   - The parameter set of each individual is passed to a simplified magnetomechanical model.
   - The model [1] computes the resulting deformation of the gripper.
   - An objective function (fitness) is evaluated.
4. **Reproduction.** Based on fitness values, individuals are selected to act as parents. Crossover is applied to produce new individuals, and mutation introduces variation in the offspring. The best individual is carried over to the next generation (elitism).
5. **Convergence check.** The algorithm evaluates whether the optimization has converged, either based on a threshold value of the objective function or after reaching a maximum number of generations. If not, the process repeats from step 3.

Although the example presented in this work involves the optimization of only three parameters, the genetic algorithm offers the advantage of flexibility in selecting which parameters to optimize. Thanks to the parametric mesh, it is also possible to optimize geometric properties such as the shape or thickness of specific elements.

Moreover, the fitness function can be easily adapted depending on the design goal. While the present study focuses on a fitness function based on the deformation of the gripper tip, it can be redefined to reflect other objectives. For example, the fitness can be defined as a function of the stress distribution in selected elements or the magnitude of reaction forces, such as the gripping force applied to an object.

In our study, the fitness function is defined as the Euclidean distance between the tip of the gripper and a predefined target point in the deformed configuration. The fitness value FF is computed as:

$$\mathrm{FF} = \left\| \boldsymbol{x}_{\mathrm{tip}}^{\mathrm{sim}} - \boldsymbol{x}_{\mathrm{target}} \right\| , \tag{3}$$

where $\boldsymbol{x}_{\mathrm{tip}}^{\mathrm{sim}}$ and $\boldsymbol{x}_{\mathrm{target}}$ are position vectors of the gripper tip from simulation and the desired target location, respectively.

## 3 Results

A case study is presented to demonstrate the capability of the proposed method for fast design. In this example, only three parameters are optimized: the magnetization directions of individual segments and the position of a permanent magnet, see Fig. 1(a). The result of the optimization, i.e., the configuration achieving the target deformation, is shown in Fig. 1(b). The entire optimization process takes only 8 minutes.

The simulated deformation is shown in Fig. 1(c). The real deformation of the gripper is measured using digital image correlation, and the final shape is shown in Fig. 1(d). Fig. 2 shows the actual soft gripper and its behavior under the influence of a magnetic field.

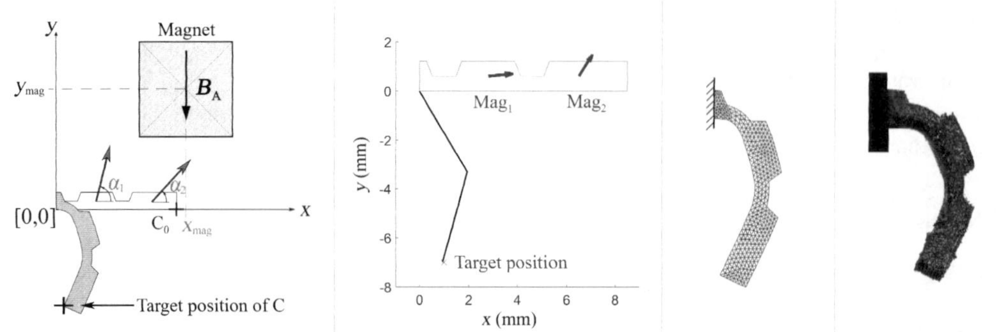


(a) optimization setup (b) optimized configuration (c) simulation (d) experiment

Fig. 1: Design of the one finger of the gripper using a simplified model and genetic algorithm.

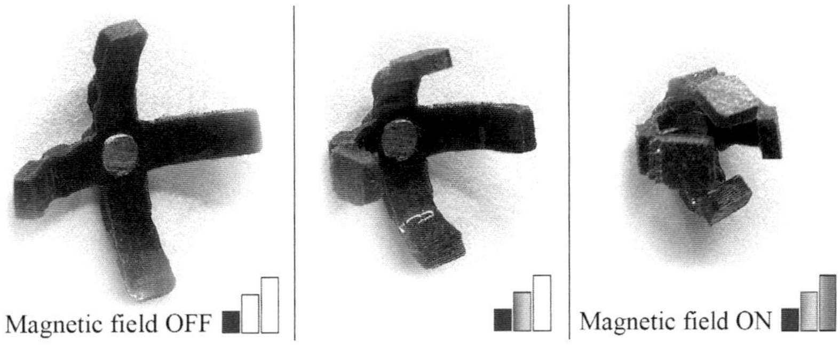


Fig. 2: Experimental response of the soft gripper to increasing magnetic field. Without a field, the gripper is open; stronger fields cause gradual finger bending and closure due to the embedded hard-magnetic particles.

The following GA parameters are used in the case study:

- Initial population size: 8 individuals
- Number of offspring per generation: 20
- Mutation range: $\pm 30\%$
- Selection: fitness-proportional

## 4 Conclusion

A simplified magnetomechanical model combined with a genetic algorithm is used for the fast design of a magnetically actuated soft gripper. The proposed framework enables efficient optimization of key design parameters and provides results within minutes. The optimized configuration achieved the desired deformation, as confirmed by both simulation and experiment. This approach demonstrates strong potential for rapid prototyping and application in soft robotic systems.

A known limitation of genetic algorithms is that they do not provide information about the global optimum of the objective function. However, in this particular application, it was observed that multiple distinct parameter combinations can satisfy the fitness criterion (i.e., achieve the desired deformation). This redundancy is beneficial, as it allows the designer to choose among feasible solutions based on additional considerations, such as manufacturability. For instance, one may prefer configurations with perpendicular magnetization directions for easier implementation.

It should be noted, however, that the simplified model has its limitations, and its accuracy is guaranteed only within a certain range of conditions. Therefore, this approach is not universally applicable to all magnetomechanical systems. Nevertheless, in the case of the soft gripper, the combination of the model and genetic algorithm proves to be highly effective, offering excellent results with minimal computational cost.

Future work may include extending the optimization to a larger set of parameters and introducing multi-objective optimization where both deformation and gripping force are considered.

## References

[1] Hermann M., et al., Simplified model of magnetomechanical coupling: simulation of magnetic-field-induced deformation of a soft gripper made of hard magnetoactive elastomer and experimental verification, Smart Materials and Structures 34 (2025) 065012, doi: 10.1088/1361-665X/ade2c2.

[2] Yang Z.,Zhang L., Magnetic actuation systems for miniature robots: A review, Advanced Intelligent Systems 2 (2020) 2000082, doi: 10.1002/aisy.202000082.

[3] Camacho J. M., Sosa V., Alternative method to calculate the magnetic field of permanent magnets with azimuthal symmetry, Revista Mexicana de Física 59 (2013) 8–17, doi: 10.31349/RevMexFisE.21.020701.

[4] Weise T., Global Optimization Algorithms: Theory and Application, 2nd Edition, Self-published, 2009.

# Computer Simulation of the Isothermal Forming Process of a Glass Element

M. Horák [1, a], M. Stará [1], G. Kehlerová [1]

[1] *Technical University of Liberec, Faculty of Mechanical Engineering*
*Department of Glass Producing Machines and Robotics*
*Studentská 1402/2, 461 17 Liberec 1, Czech Republic*
[a] *marcel.horak@tul.cz*

**Abstract:** The paper describes the problem of design and optimization of a computer model of isothermal moulding of an optical glass element by the PGM (Precision Glass Moulding) method, which takes place near the transformation temperature, when the glass gradually transitions from a solid (elastic) material phase to a viscoelastic state. This is related to the analysis of the possibilities of substituting the real material behaviour by a suitable rheological model, determination of the fundamental material constants and finally verification of the results of the computer simulation by comparison of experimentally measured data on the tensile tester. For real use in technical practice, the verified computer model was transformed in accordance with the requirements for the resulting geometry and shape of the moulded element and the process parameters of the real moulding cycle were predicted theoretically.

**Keywords:** Glass; rheological model; forming process; transformation temperature; computer model; FEM.

## 1 Introduction

The PGM method [1] is a relatively new method of manufacturing smaller optical glass parts, providing very good internal product structure, surface quality and high geometrical accuracy. The resulting production quality is influenced by the surface and material properties of the mould, the degree of wear, the shape instability during forming and cooling, the optical properties of the glass due to structural relaxation and the residual stress. Compared to standard manufacturing technologies, PGM separates the forming phase from the time-consuming heating and cooling phase. By lowering the forming temperatures, mould life is extended and, in general, heat transfer at the contact interface can be more efficiently adjusted. A discussed disadvantage is the relatively long production cycle due to the fact that the glass is formed at higher viscosities near the transformation temperature $T_g$, which defines the interface between the elastic and viscoelastic behaviour of the glass. In engineering practice, for example, three cases are commonly distinguished with respect to the change in shear modulus $G_0(T)$ as a function of temperature T:

- *the low temperature region* ($T < T_g$) where $G_0(T)$ is almost constant, the glass is in the solid phase and elastic behaviour dominates,
- *the region around the transformation temperature* ($T \cong T_g$), where $G_0(T)$ drops sharply and the glass transitions to a viscoelastic behaviour due to a decrease in the energy density of the bonds between the atoms,
- *the high temperature region* ($T > T_g$) where $G_0$ approaches zero and the glass or enamel behaves as a viscous liquid with a defined viscosity $\eta(T)$.

For temperatures up to $T_g$, the decrease in shear modulus is usually nonlinear and an approximate relationship is commonly used [2, 3, 4]:

$$G_0(T) = G_0(20°C) \cdot e^{-k \cdot (T-20)}, \quad (1)$$

where k is an empirically determined constant describing the dynamics and nature of the modulus decrease. For glass it is typically given in the range $k \cong 0{,}002 \div 0{,}005\ K^{-1}$. The actual forming process may take place during the application of a constant pressing force or, conversely, the feed rate of the forming tool for a pre-selected period of the forming cycle. Despite the fact that the applied forming procedure is seemingly simple, it is not quite easy to set the working parameters of the pressing device and for more complicated products it seems to be optimal to use a computer simulation based on the finite element method (FEM).

## 2 Computer Simulation

Two basic approaches can be used to describe the glass material behaviour during PGM forming. The first and simpler one is based on the assumption that a viscosity curve, i.e. the dependence of viscosity on temperature given by the VFT equation, is defined for the optical glass and the so-called stiff-plastic model is used. The constitutive behaviour of the glass is then described by the relation:

$$\sigma = 3 \cdot \eta(T) \cdot \dot{\varepsilon}, \quad (2)$$

where $\sigma$ is the mechanical stress and $\dot{\varepsilon}$ is the strain rate, with the possible effect of nonlinear viscous flow at higher strain rates being neglected in this case. The more complicated second approach describes the viscoelastic behaviour of glass via the general Maxwell rheological model with characteristic parallel groups of springs and dampers connected in series in accordance with Fig. 1.

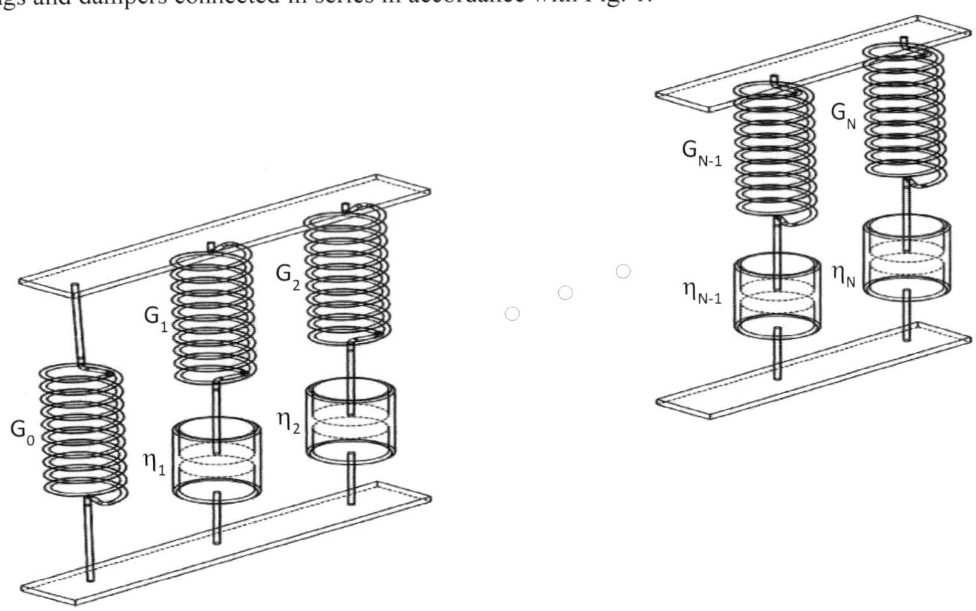


Fig. 1: General Maxwell rheological model

Maxwell model is ideal for describing the relaxation behaviour of viscoelastic materials and uses the partial relaxation times $\tau_i$ and shear moduli $G_i$ to calculate the relaxation modulus G(t) at time t in accordance with the well-known Dirichlet-Prone mathematical series [5, 6, 7]:

$$G(t) = G_\infty + \sum_{i=1}^{N} G_i \cdot e^{-\frac{t}{\tau_i}}, \quad (3)$$

where $G_\infty$ is the equilibrium shear modulus for time $t \rightarrow \infty$ s and for the case that the material will fully relax, we can write that $G_\infty = 0$ Pa. Moreover, the exponential term $e^{-t/\tau_i}$ approaches zero very quickly with increasing time t, because

$$\lim_{t\to\infty} e^{-t/\tau_i} = 0 \text{ and logic further } \lim_{t\to\infty} G(t) = G_\infty = 0. \tag{4}$$

The index i represents the i-th element in the Maxwell model and the times $\tau_i$ are characterized by the ratio of viscosity and shear moduli such that

$$\tau_i = \frac{\eta_i}{G_i}. \tag{5}$$

In the observed temperature interval in the $T_g$ region corresponding to glass forming by PGM (600 °C to 650 °C), it is relatively complicated to state whether the shear modulus $G_\infty$ can be completely neglected. It is evident that the magnitude of the transformation temperature depends on the chemical composition of the glass. The commonly used glass for optical elements is BK7 glass from the manufacturer Schott, which states that $T_g = 557$ °C. For the expected forming temperatures, the glass will exhibit rather viscoelastic behaviour with residual strength and the forming cycle will require relatively long pressing times with constant pressing force and working temperature. The problem can also be extended to structural relaxation, and the so-called Narayanaswamy model [8] has been used in glass modelling processes for many years in areas of transformation temperatures. A characteristic feature is the linearization of the problem of real glass behavior by fictitious temperatures $T_f$. Simplifying the situation, it is possible to write, on the basis of Boltzmann's superposition principle for a fictitious temperature at time t, that

$$T_f(t) = T(t) - \int_{-\infty}^{t} M_v(\xi(t) - \xi(t'))\frac{d}{dt'}(T(t'))(dt'), \tag{6}$$

where $M_v$ is the response function (sensitivity function) defined as

$$M_v(\xi(t)) = \frac{T_f - T(\infty)}{T_0 - T(\infty)}. \tag{7}$$

Software product MSC.Marc offers a parameter $\xi(t)$, the so-called reduced time, which simplifies the classical strongly nonlinear thermal rheological behaviour of glass materials to the description of a single dependence, for which the

$$\xi(t) = \int_{-\infty}^{t} \frac{\tau_{ref}}{\tau(T(t'))} dt'. \tag{8}$$

The reference relaxation time $\tau_{ref}$ of glass corresponds to the so-called $T_{ref}$ reference temperature and in general, depending on the temperature, it can be stated for the relaxation time that

$$\tau = \tau_{ref} \cdot e^{\left[-\frac{H}{R}\cdot\left(\frac{1}{T_{ref}} - \frac{x}{T} - \frac{(1-x)}{T_f}\right)\right]}, \tag{9}$$

where H is the activation energy of the process and R is the gas constant. The fractional parameter x determines the extent to which the fictitious temperature will affect the relaxation time and is in the range of values 0 to 1. Typically for the response function, using assumption (9), we can write:

$$M_v(\xi) = e^{-\frac{\xi}{\tau}} \tag{10}$$

and also for models with n relaxation times

$$M_v(\xi) = \sum_{i=1}^{n} (W_g)_i \cdot e^{-\frac{\xi}{\tau_i}}, \tag{11}$$

where $(W_g)_i$ are the weighting coefficients for which it is valid that $\sum_i^n (W_g)_i \approx 1$. The material specifications used in the simulation calculations were defined in accordance with the above theoretical assumptions and available information detailed in, e.g., [9, 10, 11].

## 2.1 Verification of Computer model

In order to verify the proper function of the computer model of the pressing process (Fig. 2), respectively the material model of the glass in the chosen temperature ranges, the results of the computer modelling were compared with the outputs from the experimental measurement of the force response on the blasting machine.

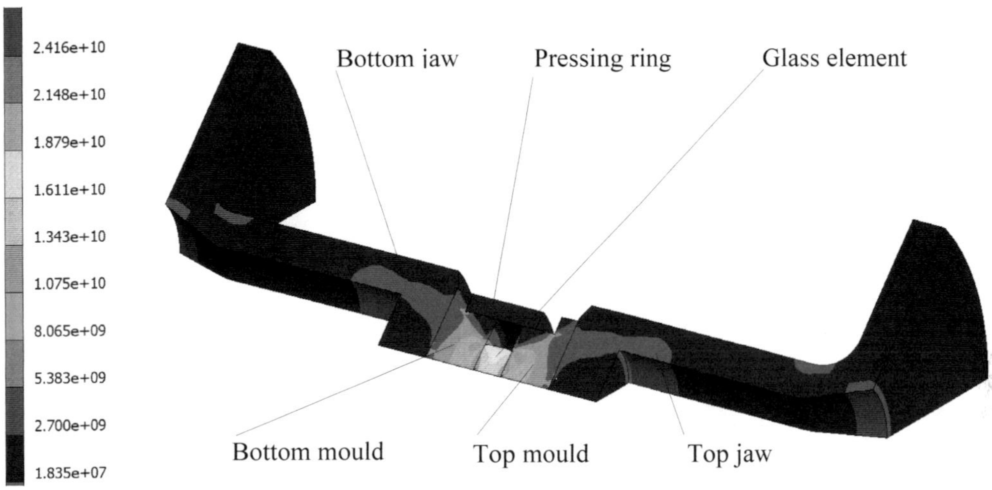


Fig. 2: Computer model (equivalent of stress [Pa])

For the measurement of the rheological response of the glass as a function of temperature, an experimental workstation was designed and commissioned in accordance with the diagram in Fig. 3, consisting of a LLOYD - instruments universal tearing machine type LR50K plus in combination with a cylindrical resistance furnace, measuring apparatus and evaluation software for the analysis of the temperature field in the space of the tested sample.

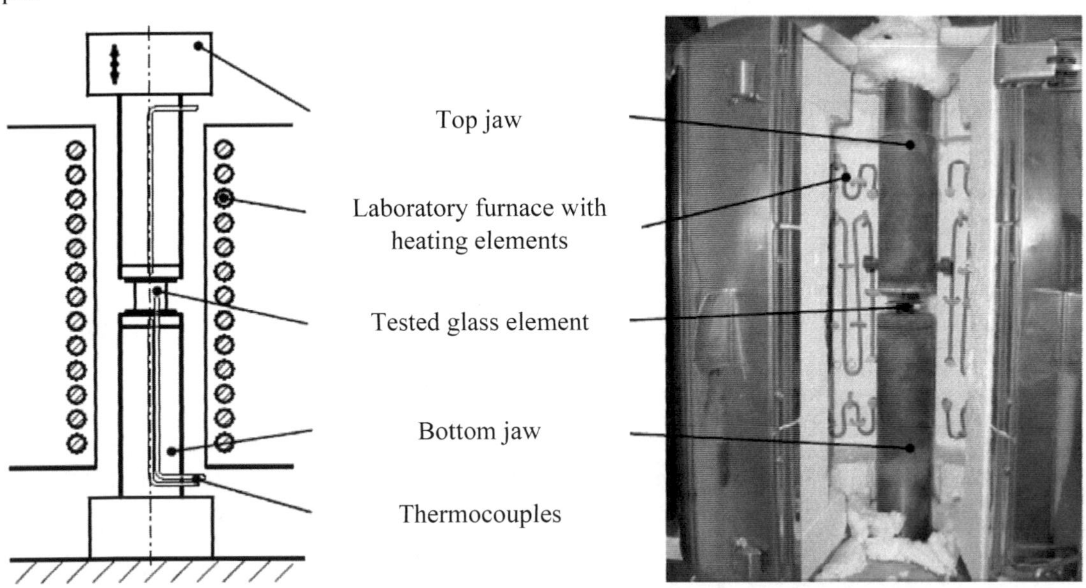


Fig. 3: Experimental stand

A series of experiments with glass cylinders with a diameter of 10 mm and a height of 10 mm were performed on the experimental stand. The force response as a function of deformation was monitored for a predefined temperature and deformation rate corresponding to the parameters of the pressing cycle. The obtained measurement results were used as a comparative data set for the optimized specification of the material constants of the rheological model. The results of the computer simulation using the optimized model showed very good agreement with the experimentally measured data. The comparison can be seen in Fig. 4.

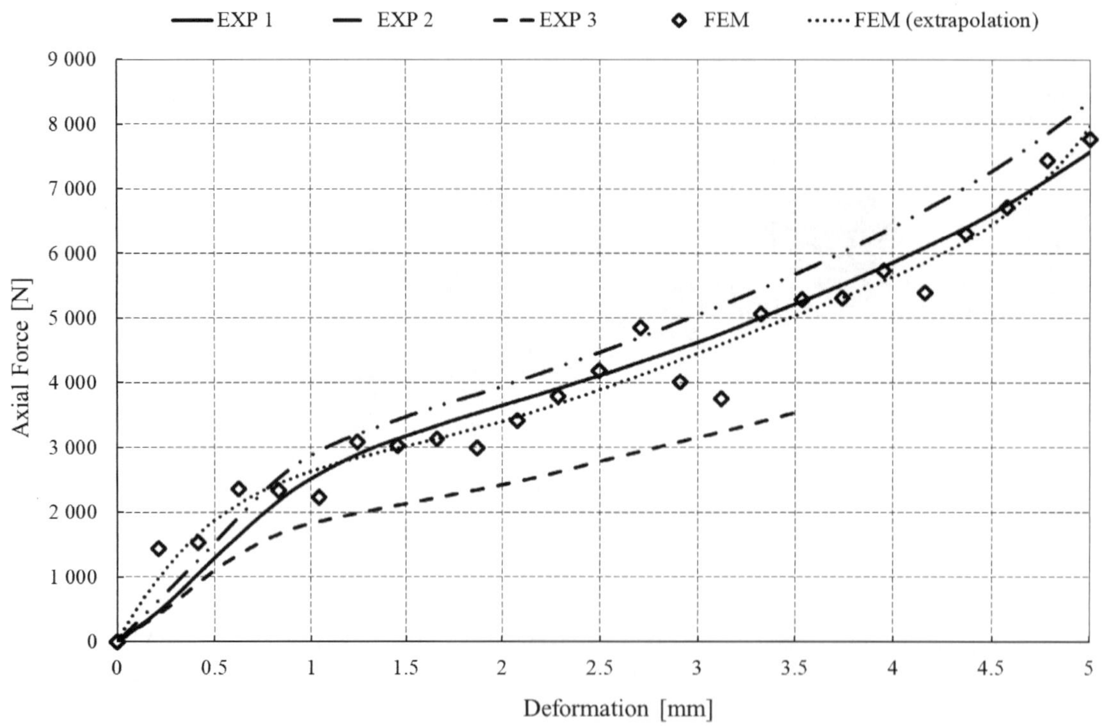


Fig. 4: Verification data

The verified model could be successfully transformed to the complicated geometry of the optical element (Fig. 5) and the simulation results were used to predict crucial parameters of the manufacturing process, such as the optimum working temperature, deformation rate and pressing force.

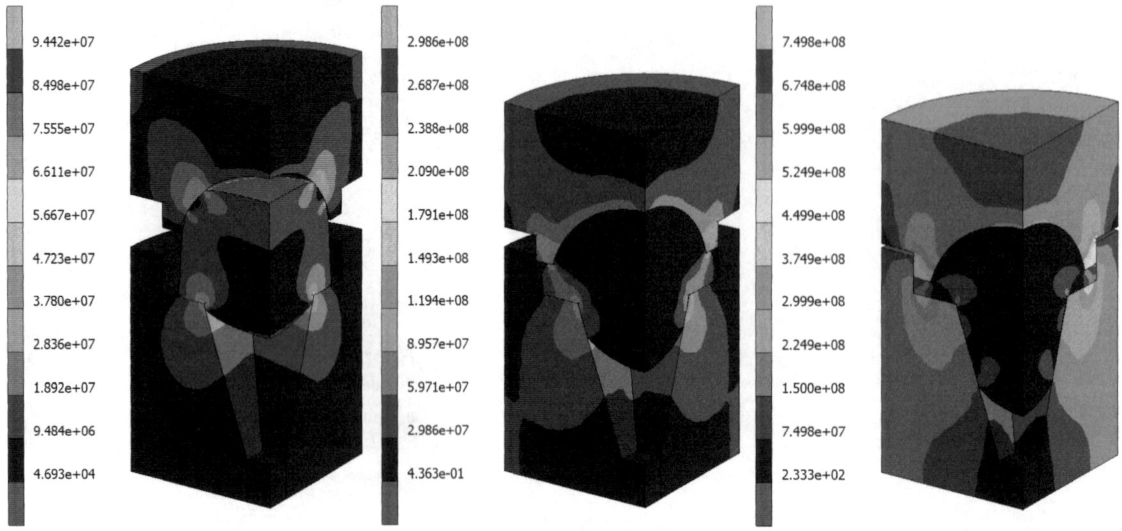


Fig. 5: Simulation of the pressing process of an optical element (equivalent of stress [Pa])

## 3 Conclusion

The present study analyses the possibilities of computer simulation of the real behaviour of optical glass in the transient temperature region. Modeling results are discussed depending on the type of rheological material model used and experimentally obtained data. The verified computer model and the proposed experimental methodology allow to solve the problem and research the process parameters of isothermal molding of shape-accurate glass products in an effective way.

## Acknowledgement

This work was supported by the Student Grant Scheme at the Technical University of Liberec through the project SGS-2023-5393.

## References

[1] ZHANG, L. and LIU, W. Precision glass molding: Toward an optimal fabrication of optical lenses. In: *Frontiers of Mechanical Engineering*. 2017, s. 3-17. ISSN 2095-0233.

[2] SHELBY, J. E. *Introduction to glass science and technology*. 2nd ed. Cambridge: Royal Society of Chemistry, c2005. ISBN 08-540-4639-9.

[3] VARSHNEYA, A. K. and MAURO, J. C. *Fundamentals of inorganic glasses*. Third edition. Amsterdam: Elsevier, [2019]. ISBN 978-0-12-816225-5.

[4] SCHOLZE, H. and LAKIN, M. J. Glass: nature, structure, and properties. Third edition. New York: Springer-Verlag, 1991. ISBN 978-1-4613-9069-5.

[5] BOUBAKER, Mohamed Bader, Benjamin LE CORRE, Yves MESHAKA and Gérard JEANDEL. Finite Element Simulation of the Slumping Process of a Glass Plate Using 3D Generalized Viscoelastic Maxwell Model. *Journal of Non-Crystalline Solids*. 2014, **405**, 45-54. DOI: 10.1016/j.jnoncrysol.2014.08.018. ISSN 00223093.

[6] NIELSEN, J.H. et al., *Finite Element Implementation of a Glass Tempering Model in Three Dimensions*. 2010, 88(17-18), 963-972. ISSN 00457949.

[7] DUFFRÈNE, L., R. et al., Temperature Dependence of the High-frequency Viscoelastic Behavior of a Soda-lime-silica Glass. *Journal of the American Ceramic Society*. 1998, **5**(81), 1278-1284. ISSN 00027820.

[8] *Marc Volume A: Theory and User Information*. 1. Newport Beach, CA: MSC Software Corporation, 2020.

[9] ZHOU, J. et al., Numerical Evaluation on the Curve Deviation of the Molded Glass Lens. *Journal of Manufacturing Science and Engineering*. 2014, **136**(5). DOI: 10.1115/1.4027342. ISSN 1087-1357.

[10] LIMBACH, R. et al., Strain-rate Sensitivity of Glasses. *Journal of Non-Crystalline Solids*. 2014, **404**, 124-134. DOI: 10.1016/j.jnoncrysol.2014.08.023. ISSN 00223093.

[11] PARSA, M.H., et al., Simulation of Windscreen Bending Using Viscoplastic Formulation. *Journal of Materials Processing Technology*. 2005, **170**(1-2), 298-303. DOI: 10.1016/j.jmatprotec.2005.05.015. ISSN 09240136.

# Determination of Rotary Joint Stiffness Coefficients using Modal Analysis

R. Huňady [1,a], T. Poštulka [1], M., Hagara [2], P. Palička [2], Z. Bobovský [1]

[1] *VSB – Technical University of Ostrava, Faculty of Mechanical Engineering, Czech Republic*
[2] *Technical University of Košice, Faculty of Mechanical Engineering, Slovakia*
[a] *robert.hunady@vsb.cz*

**Abstract:** The paper describes the methodology for determining the stiffness coefficients of a kinematic joint. The subject of investigation is a joint providing rotary motion of one arm of the 4R Bennett mechanism. The basic principle is in the FE model updating method, in which the experimentally measured natural frequencies of the analysed assembly are used to optimize the stiffness coefficients of the bushing connector representing the kinematic joint. The use of bushing with appropriately adjusted stiffness parameters significantly improves the accuracy of the numerical model compared to the model with rigid connector.

**Keywords:** FE updating, stiffness coefficients, joint, modal analysis, Bennett mechanism.

## 1 Introduction

The basic 4R Bennett mechanism [1] is one of the simplest closed-structure mechanisms. It has four links in series connected by rotary joints. In terms of the number of links and the number of kinematic pairs, it is an overconstrained mechanism that can move fully only when certain geometric and kinematic conditions are met [2]. Since this kind of mechanisms is mainly addressed in theoretical terms, it has great potential for use in pick-and-place applications [3]. However, its commercial deployment requires dealing with issues of structural rigidity, durability, accuracy, or repeatability. One of the critical steps in the design and development are static and dynamic analyses based on FEM. To obtain a correct structural response, it is necessary to have an accurate numerical model. In this paper, attention is paid to the estimation of the stiffness coefficients of one of the rotary joints of a given Bennett mechanism (Fig. 1). The FE model updating method [4] can be used for this purpose. The essence of this method is to optimize selected model parameters, e.g. material constants [5], based on minimizing the difference between numerically and experimentally obtained responses. Procedures similar to the one applied in this paper are often used to identify the joint stiffness parameters of various robotic systems and manipulators [6], [7], [8]. The differences are in the choice of design variables and the objective function.

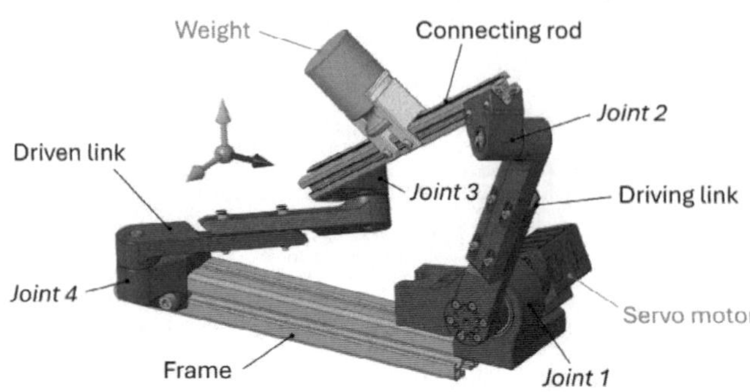


Fig. 1: 4R Bennett mechanism.

# 2 Methods and results

## 2.1 Analysed object

The object of interest is a part of Bennett mechanism. The analysed assembly consists of half of the dismountable driven link, which is attached to the frame by Joint 4, as shown in Fig. 2. The frame is made of a straight aluminium alloy profile. The plastic parts are produced by 3D printing from PETG material with a 100 % infill structure. The rotational movement of the link is ensured by two ball bearings, which are part of the Joint 4. Its detailed composition can be seen in Fig. 3a.

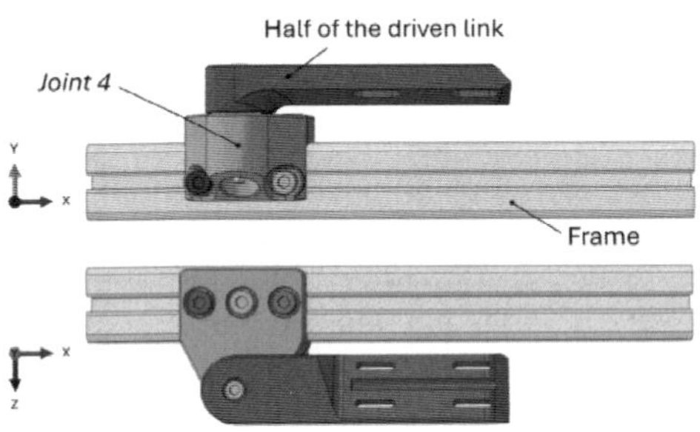


Fig. 2: Analysed assembly.

a) 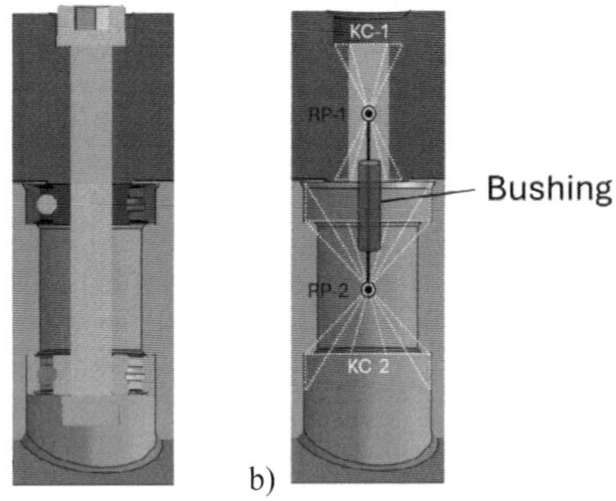 b)


Fig. 3: Sectional view of Joint 4 (a) and its FEM representation using a bushing connector (b).

## 2.2 Experimental modal analysis

Modal analysis was performed to determine the natural frequencies and mode shapes of the driven link. For this purpose, a measuring system by Hottinger Brüel & Kjær (HBK) was used. The structure was excited by a Dytran 5800 impulse hammer at 13 points and the response was measured with three-axis miniature acceleration sensor Type 4524 located at the free end of the link (Fig. 4). The measurement range was up to 1000 Hz with a resolution of 5 Hz. During measurement, the frame was firmly fixed in a heavy steel vice. To prevent the link from rotating during hammer hits, the link was supported by soft foam pads. In the given frequency range, it was possible to identify the 5 modes of the link. The first four of them are shown in Fig. 5.

a) 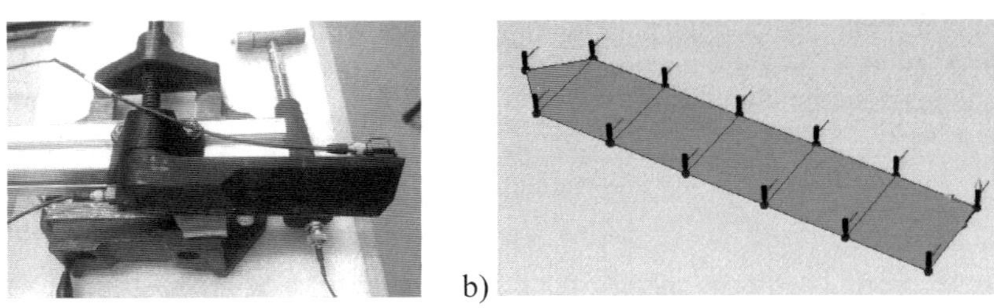 b)

Fig. 4: Experimental setup (a) and corresponding geometric model with DOFs (b).

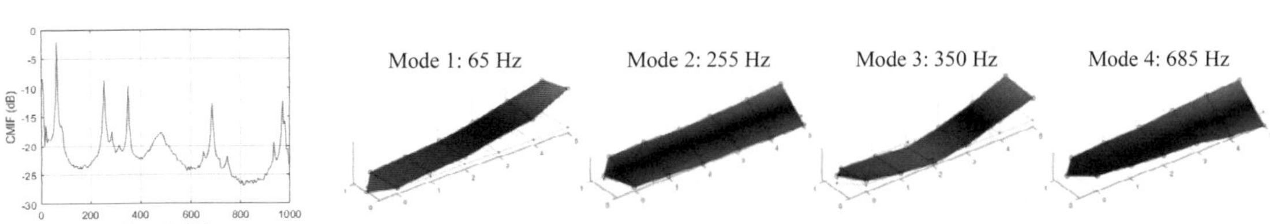


Fig. 5: Complex Mode Indicator Function and the first four mode shapes of the link.

## 2.3 Numerical modal analysis

As a next step, a simplified FE model of the analysed assembly was built in Abaqus/CAE 2024. Within the model, the bolted connections, the frame and also the bearings were neglected. The joint was modelled using a bushing connector (see Fig. 3b) with one degree of freedom allowing rotation about the joint axis. The other five degrees of freedom had a prescribed non-zero translation or torsional stiffness. The bushing was defined between two reference points (RP-1, RP-2). Each of them represents a control point that connects the contact surfaces of the corresponding component through a kinematic coupling (KC-1, KC-2). The initial stiffness coefficients that were assigned to the connector are shown in Tab. 1. Attachment of the assembly to the frame was achieved through constraints that removed degrees of freedom on the relevant contact surfaces, including those under the bolt heads and the washers. TETRA10 quadratic elements with material properties according to Tab. 2 were used to mesh the solid components. The additional weight of the accelerometer (4.8 g) was included in the form of a point mass. The final FE model of the assembly is shown in Fig. 6.

Tab. 1: Initial stiffness coefficients of the bushing.

| Stiffness coef. | K1, K2, K3 [N/mm] | K4, K5 [N.mm/rad] |
|---|---|---|
| Value | 10 000 | 10 000 |

Tab. 2: Material properties of PETG.

| Young's modulus | Poisson's ratio | Density |
|---|---|---|
| 655 MPa | 0.4 | 1080 kg/m$^3$ |

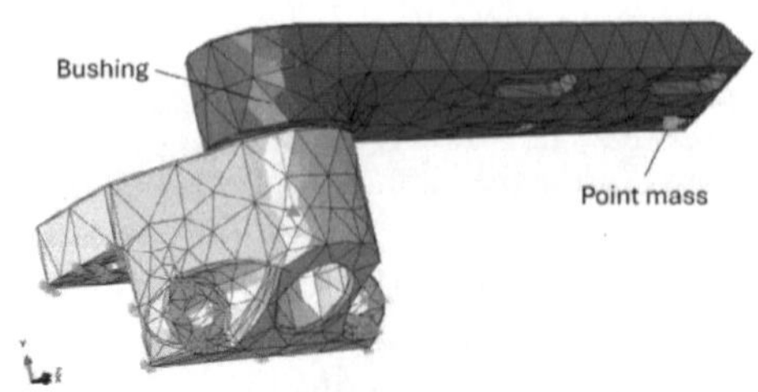


Fig. 6: FE model of the assembly.

The results of numerical modal analysis of the initial model with non-optimized stiffness coefficients are shown in Tab. 3. It is evident that the largest differences in frequency values are in the case of the first and second mode.

Tab. 3: The natural frequencies of the initial FE model.

| Mode | 1 | 2 | 3 | 4 |
|---|---|---|---|---|
| Frequency [Hz] | 26.3 | 125 | 349 | 689 |
| |  | 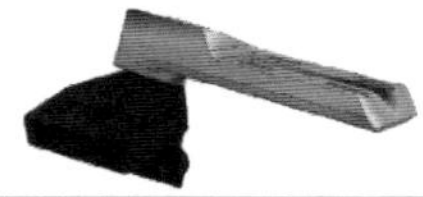 |  |  |

## 2.4 FE model updating

Estimation of the bushing parameters was performed in Isight 2024 by means of an optimization task (Fig. 7). The aim was to determine the values of the stiffness coefficients so that the natural frequencies of the FE model matched those obtained by the measurement. Only the first four natural frequencies were included in the objective function. The range for design variables was set from $10^2$ up to $10^7$. Since this was a multi-objective problem, NSGA-II (Non-dominated Sorting Genetic Algorithm) was chosen as the optimization technique. The history of objective function can be seen in Fig. 8.

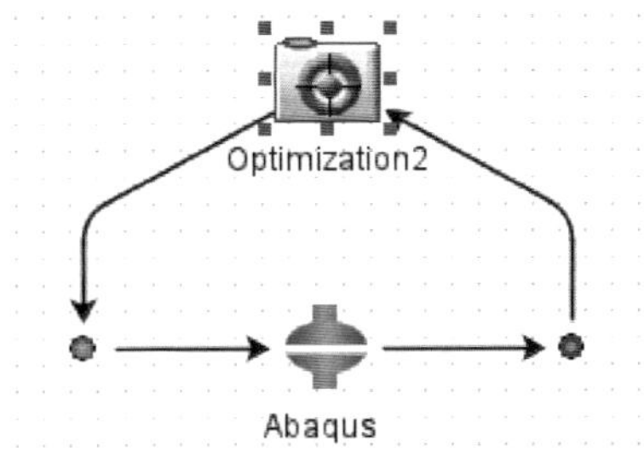


Fig. 7: Optimization loop.

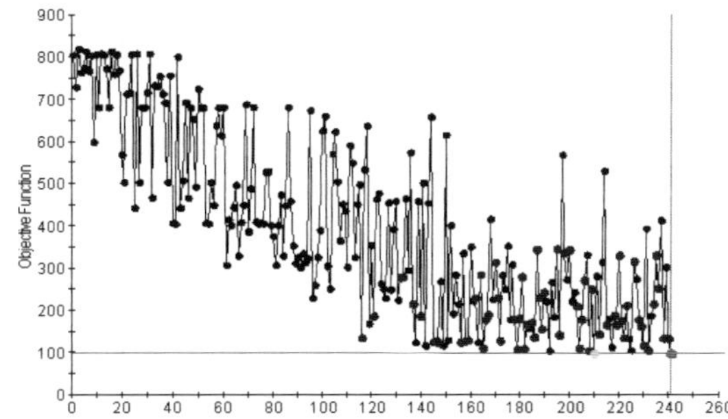


Fig. 8: History of objective function.

The results of the optimization are listed in Tab. 4. The natural frequencies and the mode shapes of the optimized FE model are shown in Tab. 5.

Tab. 4: Estimated stiffness coefficients of the joint.

| Stiffness coef. | K1 [N/mm] | K2 [N/mm] | K3 [N/mm] | K4 [N.mm/rad] | K5 [N.mm/rad] |
|---|---|---|---|---|---|
| Value | 5 236 006 | 6 485 302 | 9 096 827 | 48 098 | 67 550 |

Tab. 5: The natural frequencies of the optimized FE model.

| Mode | 1 | 2 | 3 | 4 |
|---|---|---|---|---|
| Frequency [Hz] | 62.8 | 263 | 369 | 702 |
| | 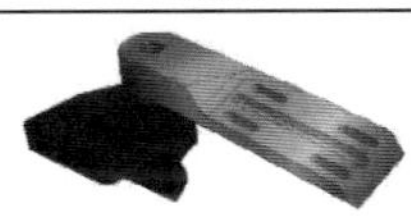 | 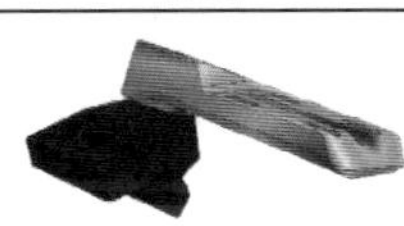 | 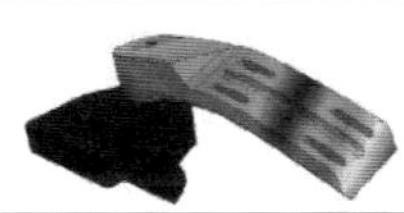 | 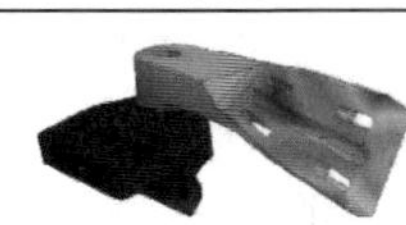 |

## 2.5 Discussion on results

Tab. 6 shows the eigenfrequency values of the FE models of the assembly with three different approaches of modelling the joint. For a better comparison, their percentage differences with respect to the values obtained by experimental modal analysis are also presented. Differences of less than 5% were considered satisfactory. It can be seen that the model with rigid rotary joint (not presented in the paper) showed the highest inaccuracy. On the contrary, the smallest differences were achieved by the model with bushing after optimization, where the maximum percentage error was +3.9%.

Tab. 6: Comparison of the natural frequencies of the FE models versus the frequencies of the real assembly.

| Mode | 1 | 2 | 3 | 4 |
|---|---|---|---|---|
| Real assembly (EMA) | 65 Hz | 255 Hz | 355 Hz | 685 Hz |
| FE model with rigid rotary joint | 119 Hz (+83%) | 486 Hz (+91%) | 591 Hz (+67%) | 789 Hz (+15%) |
| FE model with non-optimized bushing connector | 26.3 Hz (–60%) | 125 Hz (–51%) | 349 Hz (–1.7%) | 689 Hz (+0.6%) |
| FE model with optimized bushing connector | 62.8 Hz (–3.4%) | 263 Hz (+3.1%) | 369 Hz (+3.9%) | 702 Hz (+2.5%) |

## 3 Conclusion

Approximate stiffness coefficients of the rotary joint were determined using experimental modal analysis and the FE model updating method. It should be noted that the accuracy of such estimation depends not only on the geometry but also on the correct definition of the material properties of the individual parts (stiffness, mass, damping) or other entities of the model, such as contacts. However, it can be concluded that compared to an ideal rigid joint, the dynamic response of a numerical model with a finite stiffness connector is improved.

## Acknowledgement

This work was co-funded by European Union under the REFRESH–Research Excellence for REgion Sustainability and High-tech Industries project number CZ.10.03.01/00/22_003/0000048 via the Operational Programme Just Transition, by the EU NextGenerationEU through the Recovery and Resilience Plan for Slovakia under the project No. 09I03-03-V05-00015, by specific research project SP2025/042, financed by the state budget of the Czech Republic, and by the Slovak Research and Development Agency under the grant project VEGA No. 1/0152/24.

## References

[1] G. T. Bennett, The Skew Isogram Mechanism, Proceedings of the London Mathematical Society 13/1 (1914) 151-173, https://doi.org/10.1112/plms/s2-13.1.151.

[2] A. Müller, Generic mobility of rigid body mechanisms, Mechanism and Machine Theory 44 (6) (2009) 1240–1255, https://doi.org/10.1016/j.mechmachtheory.2008.08.002.

[3] T. Poštulka, et al., Applicability and Effectiveness of the Bennett Mechanism (preprint), Research Square, 2025, https://doi.org/10.21203/rs.3.rs-6296804/v1.

[4] R. Huňady, et al., Stiffness Estimation and Equivalence of Boundary Conditions in FEM Models, Applied Sciences 11 (4) (2021) 1482, https://doi.org/10.3390/app11041482.

[5] R. Huňady, et al., Determination of the Elastic Constants of a Composite Using FE Model Updating Method, in proc.: EAN 2023 - 61st Conference on Experimental Stress Analysis (2023), 75 – 81, https://www.scopus.com/record/display.uri?eid=2-s2.0-85191430815.

[6] D. Čelič, M. Boltežar, Identification of the dynamic properties of joints using frequency–response functions, Journal of Sound and Vibration 317 (1) (2008), 158-174, https://doi.org/10.1016/j.jsv.2008.03.009.

[7] W.L. Li, A new method for structural model updating and joint stiffness identification, Mechanical Systems and Signal Processing 16 (1) (2002), 155-167, https://doi.org/10.1006/mssp.2000.1339.

[8] G. Čepon, et al., On multibody-system equilibrium-point selection during joint-parameter identification: A numerical and experimental analysis, Mechanism and Machine Theory 128 (2018), 287-297, https://doi.org/10.1016/j.mechmachtheory.2018.06.006.

# Comparison of Fatigue Crack Initiation in Bonded Patches

A. Karkulín[1,a], M. Kadlec [1], B. Cabrnoch [1]

[1] *VZLU AEROSPACE Beranovych 130, 199 05 Prague, Czech Republic*

[a] *karkulin@vzlu.cz*

**Abstract:** The present research on adhesives was carried out with the aim of supporting the design of composite repairs for composite aircraft structures using a safe life approach. Skin doubler specimens of epoxy based CFRP adherents and patches were bonded using four adhesives: EC-9323, EA-9360, EA-9395 and EA-9396. The fatigue tests were carried out to determine the maximum allowable load of the bonded patches without crack initiation. A strain gauge method was successfully used. Debonding in the tapered region was further analyzed on metallographic cross sections. The fatigue results for the adhesives are unique and cannot be found in the adhesive data sheets.

**Keywords:** fatigue, failure analysis, skin doubler specimen, failure analysis.

## 1 Introduction

Composite materials have become integral to modern aerospace structures due to their high strength-to-weight ratios, corrosion resistance, and fatigue performance compared to traditional metallic materials [1]. However, damage to composite aircraft structures – whether from impacts, fatigue loading, or environmental exposure—can compromise safety, necessitating effective and certifiable repair techniques [2]. Among the various maintenance philosophies, the safe-life approach remains a foundational strategy in both design and repair, particularly where damage tolerance and fail-safety cannot be guaranteed.

To assess the fatigue performance of repaired composite structures, coupon-level testing is a fundamental step in the certification process and is put into perspective in Fig. 1a showing the test pyramid for certification of adhesive repair for composite laminate (as is outlined in [3]). A particularly effective method involves the use of Skin Doubler Specimens (SDS), which replicate bonded patch repairs and enable monitoring of the health of adhesive joint by the means of strain gauges. An example of typical external patch repair can be seen in Fig. 1b. These tests are designed to simulate service-like loading and can highlight issues such as disbond initiation and propagation under cyclic loads.

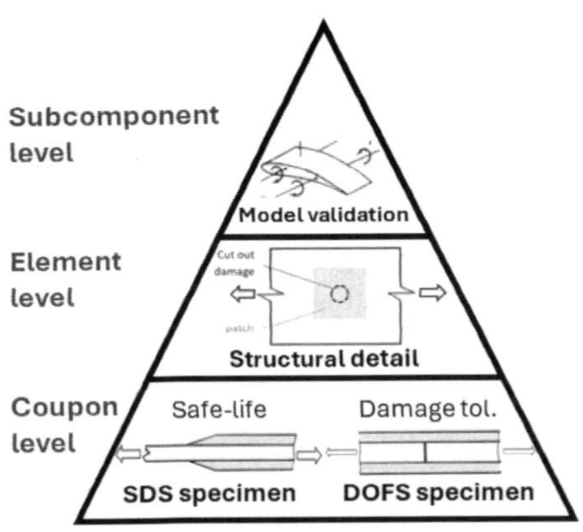


a) Certification test pyramid of adhesive repair for a composite laminate.

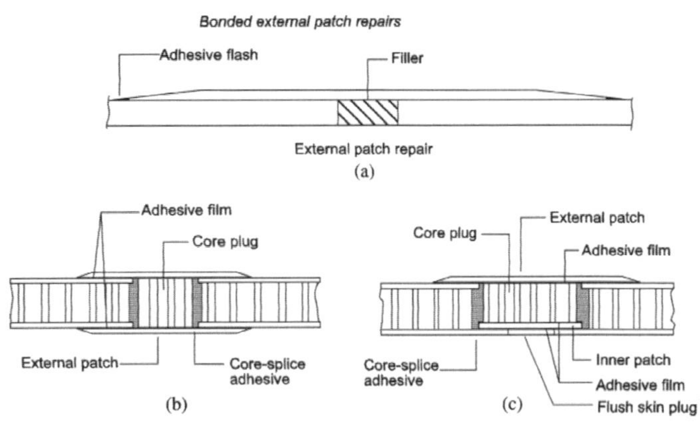


b) Typical external patch repairs illustrating repairs of laminates and sandwiches [3].

Fig. 1: The repair of composite structures – from coupons to certified bonded patch repairs.

A significant consideration in safe-life fatigue design and repair is the role of knockdown factors – conditions and material variables that reduce the ideal performance of composites in service. These include humidity, elevated or cycling temperatures, ply orientation, void content, fiber waviness, and thickness variations, all of which can accelerate fatigue damage or reduce bondline integrity in repairs [4,5,6]. Environmental factors such as moisture ingress can reduce interlaminar shear strength and lead to matrix softening, while improper ply stacking sequences can create local stress concentrations, affecting both static and fatigue properties. The effects of environment (temperature and humidity conditioning) have significant effect on type of failure mode (delamination, fiber tear or adhesive failure) and on shear strength of adhesive bond [7].

An essential element of this development is the selection and qualification of structural adhesives used in bonded repairs. The presented research targeted several such adhesives to support the design of composite repairs for composite aircraft structures [8]. The main form of repair is through adhesively bonded composite and for this reason, SDS test specimens were designed based on [9,10]. Main damaging strains on adhesive joints are shear and peel. The SDS specimens are used to evaluate fatigue threshold and are representative of the outer, peripheral part of the patch evaluated under safe life philosophy.

The fatigue test results were preceded by thorough static testing of Double Lap Shear Strength specimens [11,12] and are displayed side by side in the result section of presented research.

## 2 Material and specimen

The Skin Doubler Specimens (SDS) test specimens are composed of several parts: adherent that makes the main body of specimen, patch that is applied onto the adherent and tabs that are each side of specimen used for clamping. The exact geometry and location of strain gauges (SGs) of SDS is shown in Fig. 2. The thickness of adherent was 3 mm and thickness of adherent with 2 patches was 6 mm. The adherent was manufactured from HexPly 8552 resin reinforced by AS4 Carbon Fiber Tape while the patch used EA-9396 reinforced by ECC 450-5 Aero Carbon Fiber Tape. The specimens were made in 4 batches for various types of adhesives: EC-9323, EA-9360, EA-9395 and EA-9396.

SGs had to be carefully selected for composite specimens and in the end CEA-13-125UW-350 SGs with gage factor 2.145 were chosen for monitoring of strain. The wider size of SGs and higher resistance is desirable due to the many factors like: heterogeneity (averaging over larger area which can contain local anomalies), increased stability / reduced noise and heat dispersion (metallic specimen conduct and disperse heat better).

Furthermore, contrast paint was applied using white marker (Edding 790 paint marker) to the edges of the SDS specimens so the crack can be documented after fatigue test. The paint was applied in one or, at maximum, two passes. Three or more passes runs risk of hiding the crack beneath the layer of paint.

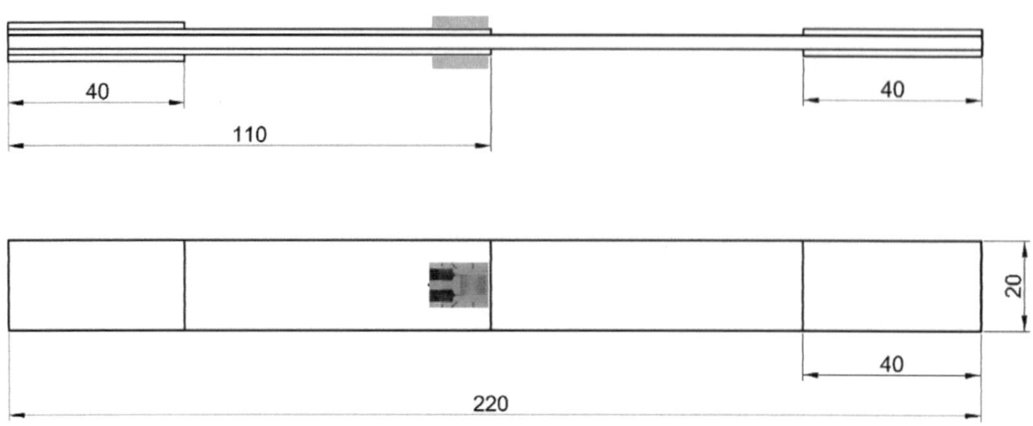


Fig. 2: The geometry of Skin Doubler Specimens for fatigue tests.

# 3 Fatigue procedure

The specimen was clamped with 40 mm clamping length on both sides. The loading was set to harmonic (sinusoidal shape) with constant amplitude load, coefficient of loading cycle asymmetry R = 0.1 and loading frequency of 6 Hz. The test at loading level is performed until fatigue threshold is reached (strain drop ≥10%) which leads to test termination, or 50 000 cycles is reached, and test is repeated at higher loading level. The fatigue procedure is displayed on Fig. 3 where the disbonding occurs at the root of adhesive joint (Fig. 3a) connecting adherent to the patch which results in drop of strain and termination of test. The SG data was filtered using peak detection algorithm to plot strain amplitude against the number of cycle (Fig. 3b).

The test setup of fatigue test of SDS specimen instrumented with SGs and with white contrast paint can be observed in Fig. 4.

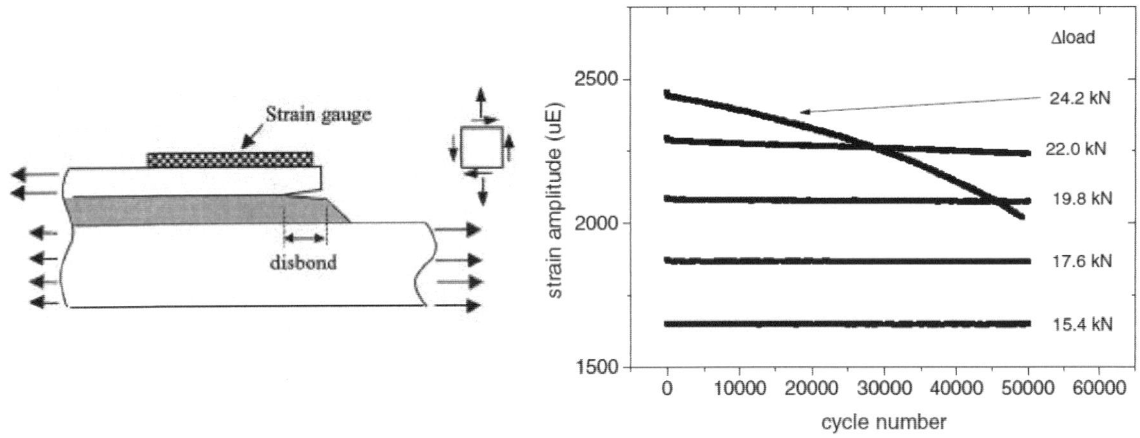


a) Root of adhesive joint with disbanding. b) Strain gauge reading at different load levels.

Fig. 3: Test procedure overview [9].

Fig. 4: Test setup of SDS instrumented with SGs and with white contrast paint.

# 4 Results and discussion

In total four sets of adhesives were tested (5 test samples per adhesive). The resulting mean value ± standard deviation of load level without strain drop are displayed in tabular form in Tab. 1 and in graphical form in Fig. 5a where the results fatigue results are compared against static Double Lap Shear Strength (DLSS) test results ([11] for EA-9395 & EC-9323 and [12] for EA-9396 and EA-9360). An example of strain measurement (for sample ID 9360-2) is documented Fig. 5b.

When compared against static DLSS test results the fatigue results do not match exactly. The EC-9323 adhesive had the highest shear strength with EA-9360, which had best performance in fatigue tests, having lower mean shear strength with rather high standard deviation. Furthermore, the maximum load without strain drop for EA-9395 is significantly higher than the static shear strength would suggest. Moreover, the scatter of data (standard deviation) is low when compared to EA-9360 where the first discrepancy between shear strength and fatigue occurred.

Tab. 1: The results of fatigue tests for SDS specimens.

| Adhesive | Load level without strain drop (kN) |
|---|---|
| EA-9395 | 5.8 ± 0.45 |
| EC-9323 | 6.8 ± 0.57 |
| EA-9396 | 4.5 ± 0.61 |
| EA-9360 | 7.2 ± 0.57 |

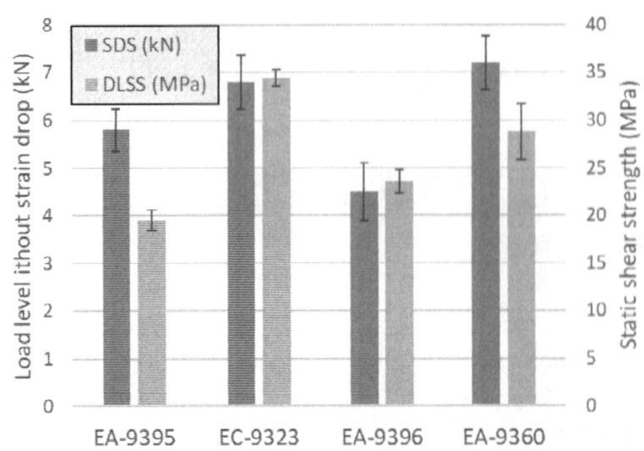


a) Maximum load without strain drop results for all tested adhesives compared against DLSS tests [11,12].

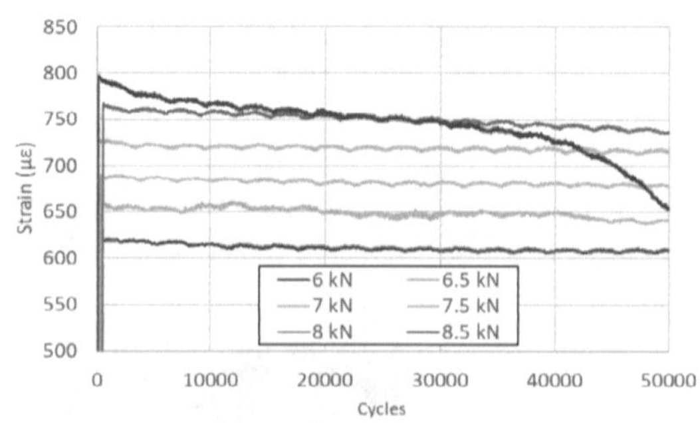


b) The peak strain recording including eventual strain drop at 8.5 kN for sample ID 9360-2.

Fig. 5: The results of fatigue tests.

The failure analysis was performed in the region at the root of adhesive joint connecting adherent to patch. In total, specimens underwent three separate inspections under microscope: the initial inspection of white contrast paint and more thorough analysis in two different planes: the plane of symmetry (plane of strain gauge) and the plane at the edge of specimen. An example of an initial inspection is shown in Fig. 6 where an overview is shown with two highlighted regions and two blue lines that outline adhesive separating adherent from patch. Note that gamma correction filter with gamma coefficient of 1/4 was used to show crack. The regions of interest displayed in Fig. 6 are shown in greater detail in Fig. 7a, a region containing crack, and in Fig. 7b, a region containing shear tears angled at 45° in the direction of shear loading.

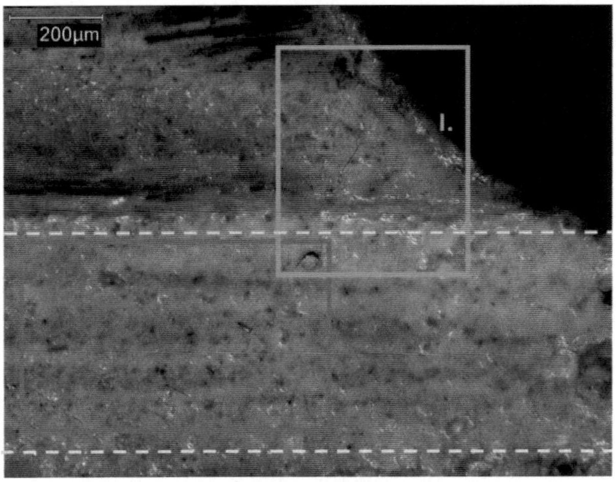


Fig. 6: The initial inspection of the specimen under microscope containing: region no. I. displayed in Fig. 7a, region no. II. Displayed in Fig. 7b and two blude dashed lines that outline adhesive layer separating adherent and patch.

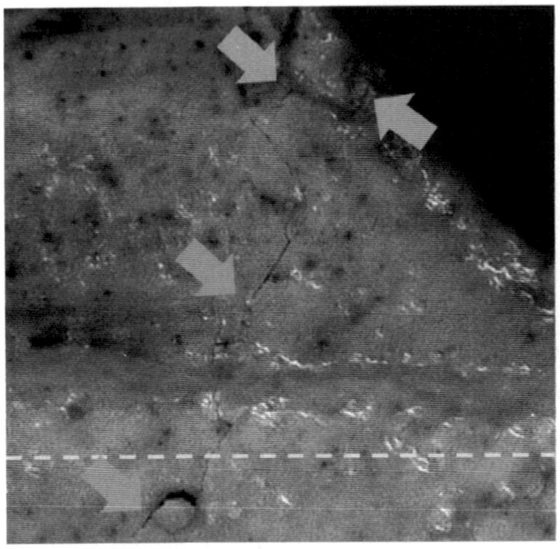

a) Detail of region no. I. containing crack with green arrows indicating its position and blue dashed line outlining upper line separating adhesive layer from patch

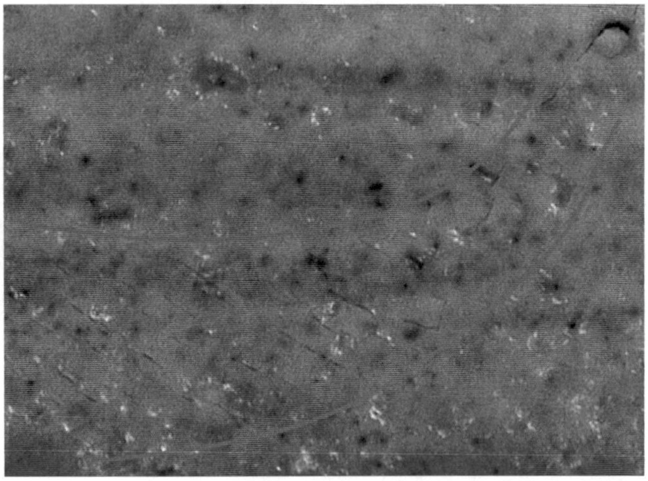

b) Detail of region no. II with two dark orange lines that create boundary for area containing 45° shear tears in the layer of adhesive between adherent and patch.

Fig. 7: The initial inspection of specimen under microscope with highlighted crack / indication of crack.

The thorough analysis shows three main failures in the adhesives. The three failures, see Fig. 8, are characterized by: type I failure follows the edge of the patch to its corner and then further towards adherent; type II failure misses the patch and stops at or near the adherent; type III failure starts further away from the patch, reaches the adherent and then follows the adherent.

The examples of failure I in the plane of strain gauge (symmetry) and the plane at the edge of specimen is shown in Fig. 9a and Fig. 9b. respectively (note that the initial inspection showed in Fig. 6, Fig.7a and Fig. 7b is of the same specimen and same location). Failure II and failure III are shown in Fig. 10a and Fig. 10b respectively. Cracks were highlighted by orange arrows.

The patch and adherent regions, where the direction of fibers is perpendicular to the direction of failure growth, proved to be susceptible to the failure propagation. In the example of type I failure (Fig. 9a and Fig. 9b) the disbonding also enters the patch region in the corner near the root of adhesive joint. Similarly, in the type III failure (Fig. 10b) the disbanding enters the top layer of adherent (the top two layers are perpendicular to the direction of failure propagation) before the test is terminated.

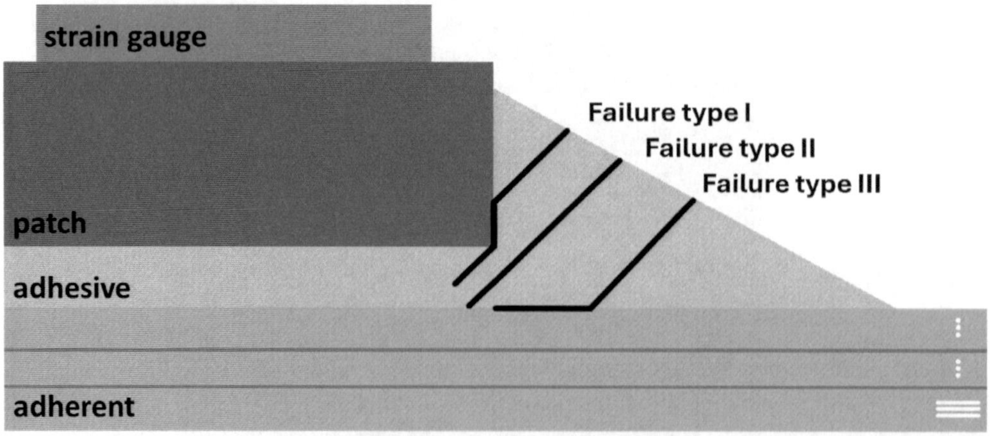


Fig. 8: Observed failures at the root of adhesive joint connection between adherent and patch.

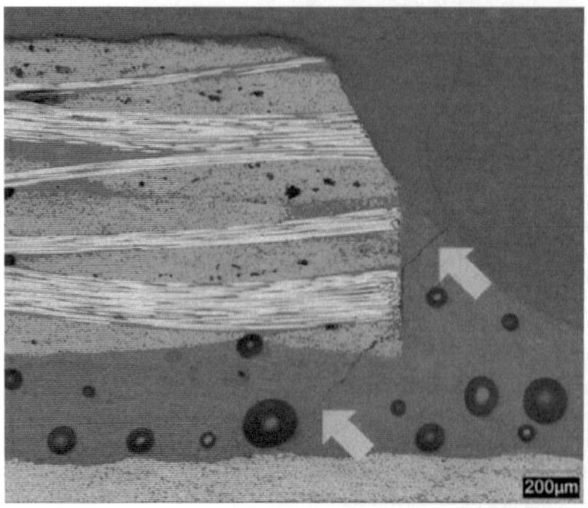


a) Plane of strain gauge (symmetry) with orange arrows indicating position of the crack.

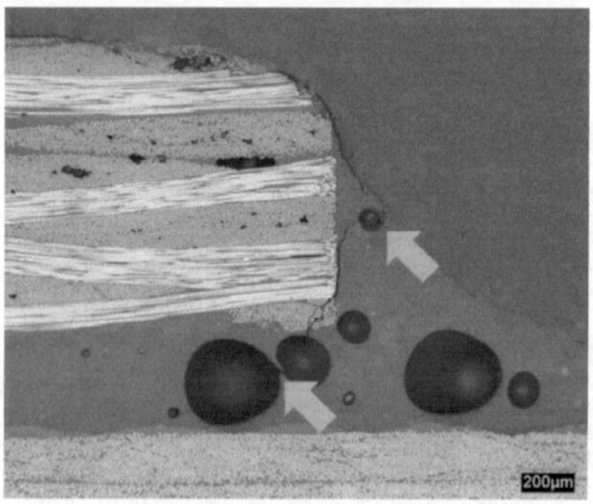


b) Plane at the edge of specimen with orange arrows indicating position of the crack.

Fig. 9: The examples of type I failure following the edge of the patch to its corner and then further towards adherent in two different planes of SDS specimen (same specimen and same location as in Fig. 6, Fig. 7a and Fig. 7b).

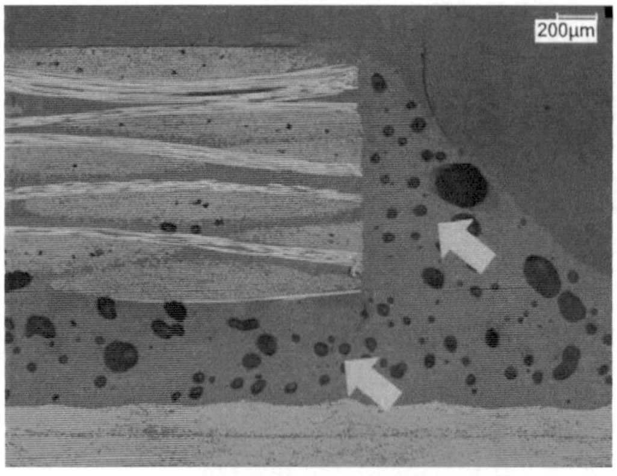


a) Type II failure that misses the patch and stops at or near the adherent with orange arrows indicating its position.

b) Type III failure that starts further away from the patch, reaches the adherent and then follows the adherent with orange arrows indicating its position.

Fig. 10: The examples of failures in the plane of strain gauge.

## 5 Conclusion

The fatigue testing of SDS samples using different adhesives provided valuable insight into their relative performance under cyclic loading conditions. Among the used adhesives the EA-9360 reached the highest maximum load level at the crack initiation. The EC-9323 was within one standard deviation below EA-9360 with other adhesives having poorer performance. Failure analysis was performed at two cross-sections and three types of failure in adhesive were identified. The adhesive and composite near the root of adhesive with orientation perpendicular to the direction of crack propagation were most susceptible to crack growth.

## Acknowledgement

The work presented in this report is co-financed with the state support from the Technology Agency of the Czech Republic under the National Centres of Competence, project national Competence Centre for Aeronautics and Space, sub-project Modern manufacturing Processes under grant agreement No TN02000009/05.

## References

[1] P. Alam, et al., The fatigue of carbon fibre reinforced plastics - A review, Composites part b-Engineering (2019), vol. 166, pp. 555-579. https://doi.org/10.1016/j.compositesb.2019.02.016

[2] K. B. Katnam, L. F. M. Da Silva, T. M. Young, Bonded repair of composite aircraft structures: A review of scientific challenges and opportunities. Progress in Aerospace Sciences (2013). 61. 26-42. 10.1016/j.paerosci.2013.03.003.

[3] U.S. Department of Transportation - Federal Aviation Administration (FAA), Advisory Circular – Composite Aircraft Structure, AC20-107B, FAA, 2010.

[4] C. E. Harris, J. H. Starnes Jr., M. J. Shuart. Advanced Durability and Damage Tolerance Design and Analysis Methods for Composite Structures—Lessons Learned From NASA Technology Development Programs (NASA/TM-2003-212420). NASA Langley Research Center, 2003.

[5] S. Budhe, M. D. Banea, S. de Barros. Bonded repair of composite structures in aerospace application: a review on environmental issues. Applied Adhesion Science (2018). https://doi.org/10.1186/s40563-018-0104-5

[6] S. N. Nair, A. Dasari, C. Y. Yue, S. Narasimalu. Failure Behavior of Unidirectional Composites under Compression Loading: Effect of Fiber Waviness. Materials 2017, 10, 909. https://doi.org/10.3390/ma10080909

[7] M. Kadlec; B. Cabrnoch, R. Hron, Effect of Extreme Environments on Adhesive Joint Performance. J. Compos. Sci. 2024, 8, 511. https://doi.org/10.3390/jcs8120511

[8] A. Baker, Repair of metallic airframe components using fibre-reinforced polymer (FRP) composites. In Rehabilitation of Metallic Civil Infrastructure Using Fiber Reinforced Polymer (FRP) Composites, 2014.

[9] A. Baker, Proposed Approach for Certification of Bonded Composite Repairs to Flight-Critical Airframe Structure, Appl Compos Mater 18 (2011) 337–369.

[10] A. Baker, Development of a Hard-Patch Approach for Scarf Repair of Composite Structure, Air Vehicles Division, Defence Science and Technology Organisation, DSTO-TR-1892.

[11] M. Kadlec, "Materiálové zkoušky NCK WP2 DLSS.", Laboratory of materials, report no. P-DAV-ZLM-013-24, Praha – Letňany, Czech republic, April 2024.

[12] M. Kadlec, "Materiálové zkoušky DLSS lepidel EA-9360 A EA-9396.", Laboratory of materials, report no. P-DAV-ZLM-046-24, Praha – Letňany, Czech republic, September 2024.

# Evaluating Poisson's Ratio in Conventional and Auxetic Cellular Structures

J. Kostka [1,a], B. Schürger [1], P. Frankovský [1], P. Palička [1], M. Hagara [1]

[1] *Department of Applied Mechanics and Mechanical Engineering, Faculty of Mechanical Engineering, Technical University of Košice, Letná 1/9, 042 00 Košice-Sever, Slovakia*
[a] *jan.kostka@tuke.sk*

**Abstract:** This work investigates and compares the mechanical characteristics of conventional and auxetic (re-entrant) structures, with a focus on their Poisson's ratio behavior under mechanical loading. Analytical calculations, finite element simulations in SolidWorks, and experimental tests using 3D-printed PLA models were performed. The study demonstrates the auxetic behavior—specifically, a negative Poisson's ratio—of the re-entrant structure and compares it with a conventional honeycomb model. Differences in deformation patterns were evaluated using a camera-based measurement system and MATLAB post-processing. Results from analytical, numerical, and experimental approaches are discussed, highlighting the accuracy and limitations of each method.

**Keywords:** auxetic structures; Poisson´s ratio; FEA; 3D printing; mechanical testing.

## 1. Introduction

Poisson's ratio is a key mechanical property that describes how a material deforms under load, defined as the ratio of transverse to longitudinal strain. For most conventional materials, this value lies between 0 and 0.5, indicating lateral contraction when stretched and lateral expansion when compressed (Fig. 1). In contrast, auxetic materials, characterized by a negative Poisson's ratio (NPR), exhibit the opposite behavior—they expand laterally when stretched and contract when compressed (Fig. 2) [1]. This counterintuitive deformation response arises not from the base material itself, but from engineered changes in the geometry or topology of cellular structures. One particularly important advantage is their enhanced resistance to indentation. Under compressive loads, auxetic structures densify locally in the impact region, absorbing energy more efficiently than traditional materials (Fig. 3) [2].

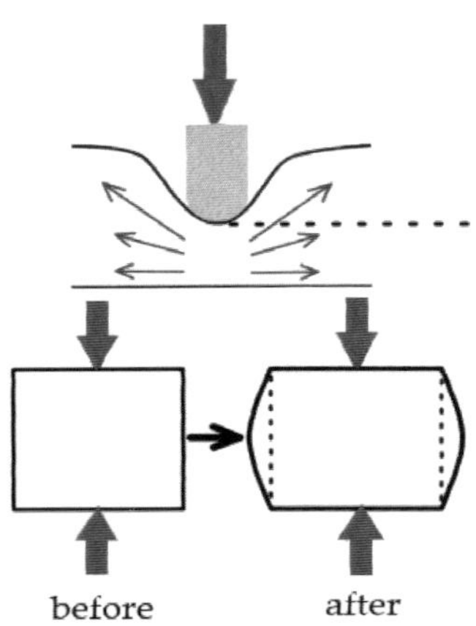


Fig. 1: Deformation of a Conventional Material Under Compression (Lateral Expansion).

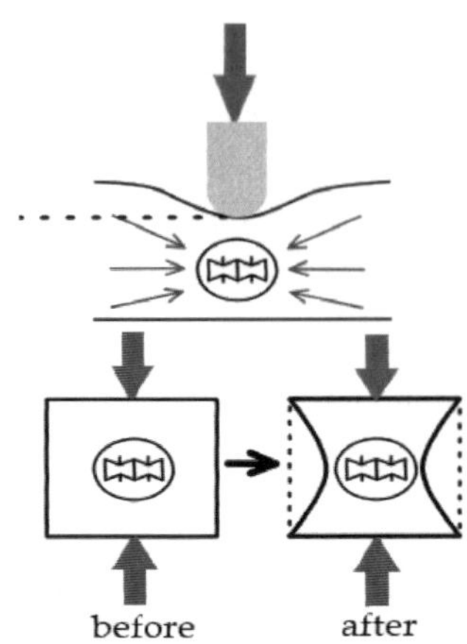


Fig. 2: Deformation of an Auxetic Material Under Compression (Lateral Contraction).

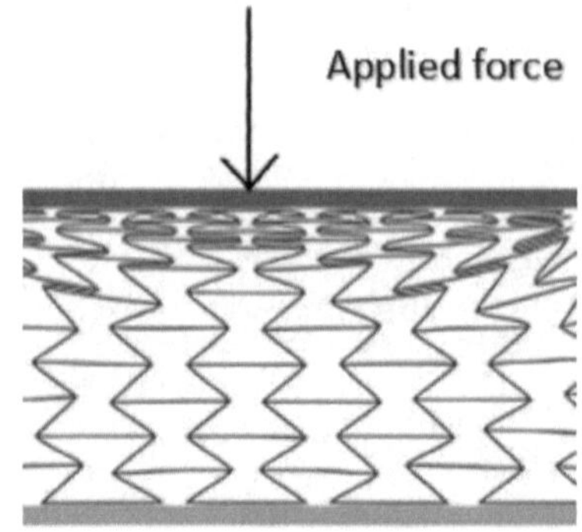


Fig. 3: Local Densification of an Auxetic Structure Under Applied Force.

These factors—alongside microstructure, manufacturing process, and loading direction—play a critical role in determining the mechanical performance of materials. This paper explores and compares these deformation behaviors by examining different structural geometries under axial loading.

Auxetic materials possess several unique properties that distinguish them from traditional counterparts. These include synclastic curvature, tunable permeability, and superior indentation resistance. Such properties stem primarily from the geometry of the structure rather than the intrinsic material composition [3-7]. Unlike conventional materials, which form saddle-shaped surfaces under bending, auxetic structures adopt dome-like forms, making them suitable for complex geometries in fields such as biomedical engineering and automotive design [8-11].

Their porous microstructures respond dynamically to stress, enhancing their performance in filtration applications [12]. Additionally, auxetic materials demonstrate superior fatigue resistance, energy absorption, and fracture toughness—even with low density and stiffness. This makes them ideal for lightweight and multifunctional applications such as sandwich panels [13], adaptive filters [14], protective gear [15], textiles [16] and acoustic dampers [17].

## 2. Materials and Methods

This study focuses on the mechanical characterization of two cellular structures: a conventional honeycomb model and a re-entrant auxetic honeycomb model (Fig. 4).

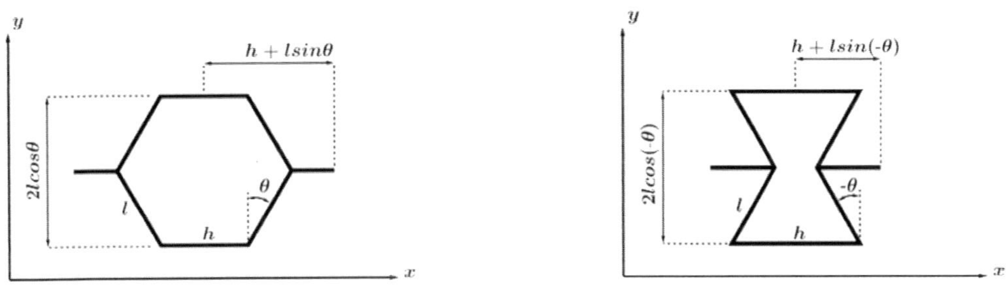


a) Conventional Honeycomb Model b) Re-entrant Honeycomb (Auxetic) Model

Fig. 4: Unit Cell Geometries of the Analyzed Structures.

Both models are designed as periodic structures composed of 11 columns and 10 rows of unit cells, with a uniform thickness of 2 mm. Their overall dimensions and unit cell configurations are presented in Fig. 12

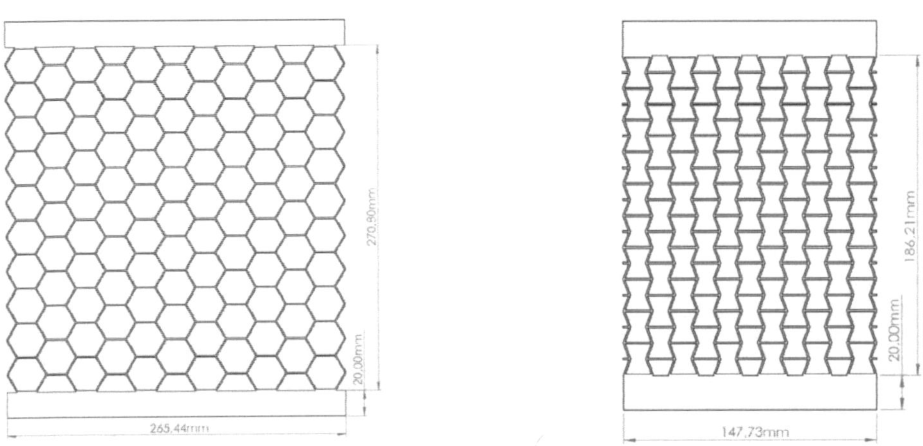


a) Conventional Honeycomb Model b) Re-entrant Honeycomb (Auxetic) Model

Fig. 5: Overall Dimensions of the Specimens Used for Analysis.

To evaluate the deformation behavior and calculate the Poisson's ratio, a multi-method approach was employed, combining analytical modeling, finite element simulations, and experimental testing.

In the **analytical phase**, closed-form expressions derived from geometric parameters of the unit cells were used to estimate Poisson's ratio for both configurations. The conventional model, based on regular hexagons, is expected to exhibit positive Poisson's ratios, while the re-entrant design—with internal angles exceeding 180°—is designed to exhibit negative values.

The **numerical analysis** was conducted using the Finite Element Method (FEM) to capture deformation mechanisms more accurately, including nonlinear effects and structural anisotropy. Simulations provided detailed strain distributions and numerical estimates of Poisson's ratio under uniaxial loading.

In the **experimental phase**, both the conventional and re-entrant structures were fabricated using PLA material and 3D printing technology. The samples were mounted in a tensile testing machine, and their deformation was recorded using a Dantec Dynamics high-speed camera. To calculate Poisson's ratio, white reference markers were applied to the surface, and their displacement was analyzed using a custom MATLAB script.

## 2.1 Analytical Determination of Poisson's Ratio

The geometric dimensions of the unit cells used for the analytical calculation are shown in Figure 6. The conventional honeycomb unit cell (Fig. 6a) has a height of 15 mm, side length of 15 mm, and an internal angle of 120°. The cell wall is inclined by 30° from the vertical axis, resulting in an inner height of 25.98 mm and wall thickness of 1 mm. The re-entrant honeycomb unit cell (Fig. 6b) also has a height of 15 mm, with inclined wall segments of 9.32 mm, an internal angle of 70°, an inner width of 17.51 mm, and a wall thickness of 1 mm.

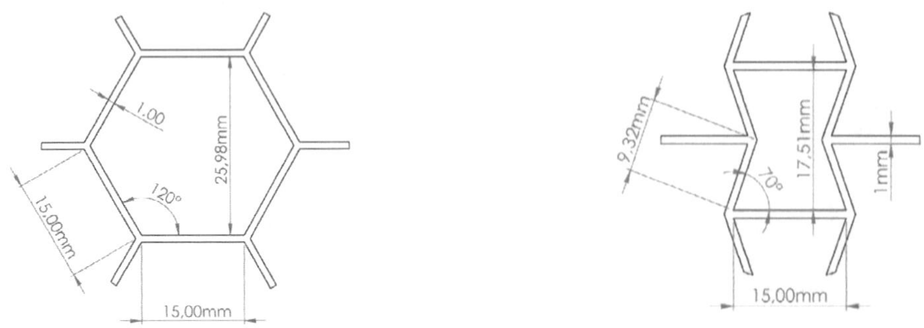


a) Conventional Honeycomb Model b) Re-entrant Honeycomb (Auxetic) Model

Fig. 6: Unit Cell Dimensions.

Based on these dimensions, the analytical values of Poisson's ratio were determined using the following equations:

- For the conventional honeycomb model, Poisson's ratio was calculated using Eq. (1):

$$\nu = \frac{(cos\theta)^2}{(h/l+sin\theta)sin\theta} = \frac{(\cos(30))^2}{\left(\frac{15}{15}+\sin(30)\right)sin30} = 1 \tag{1}$$

- For the re-entrant honeycomb (auxetic) model, Eq. (2) was used:

$$\nu = -\frac{2(1-cos\theta)}{sin\theta} = \frac{2(1-\cos(70))}{\sin(70)} = \text{-}1{,}4 \tag{2}$$

## 2.2 Numerical Determination of Poisson's Ratio

Numerical analysis was performed to determine the Poisson's ratio of both the conventional and re-entrant auxetic honeycomb structures. The goal was to evaluate both global and local deformation behavior under axial tensile loading.

The numerical simulations were carried out using SolidWorks Simulation, applying a static linear analysis. The bottom edge of the structure was fully fixed, while a tensile force of F = 100 N was applied to the top edge. After the simulation, maximum displacements in both X and Y directions were recorded (Fig.7 and Fig. 8).

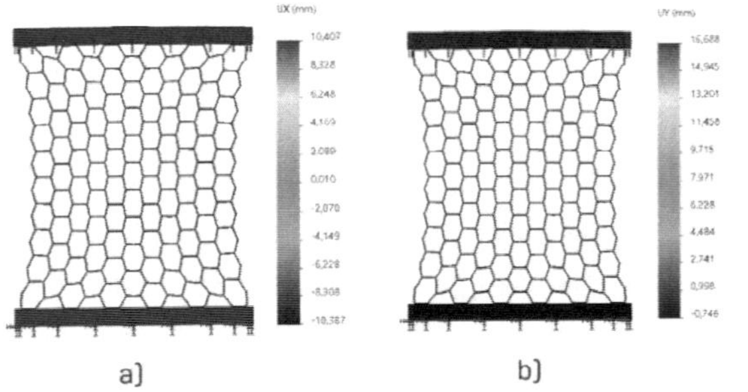


Fig. 7: Directional Deformation of Conventional Structure: A) X-Direction; B) Y-Direction.

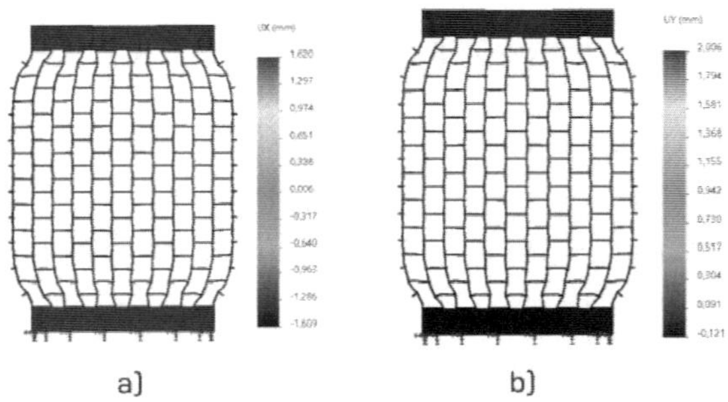


Fig. 8: Directional Deformation of Auxetic Structure: A) X-Direction; B) Y-Direction.

Poisson's ratio was then calculated using the relation:

$$\nu = -\frac{\varepsilon_y}{\varepsilon_x} \tag{3}$$

, which for the conventional model, the computed Poisson's ratio was $\nu$= 0,73 and for the re-entrant auxetic model, the Poisson's ratio was $\nu$= -1,24.

To provide a more detailed insight, Poisson's ratio was also evaluated based on displacements at specific points on the structure. Virtual sensors were placed at points 1, 2, 3, and 4 (Fig. 9), and displacements were measured before and after deformation using SolidWorks sensor functions (Fig.10).

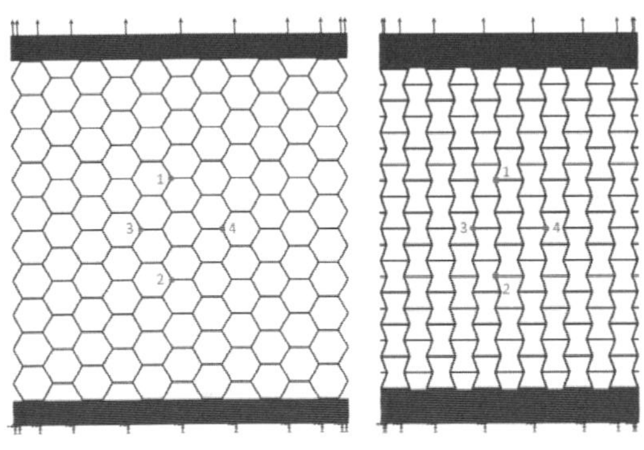

Fig. 9: Deformation of a Conventional Material Under Compression (Lateral Expansion).

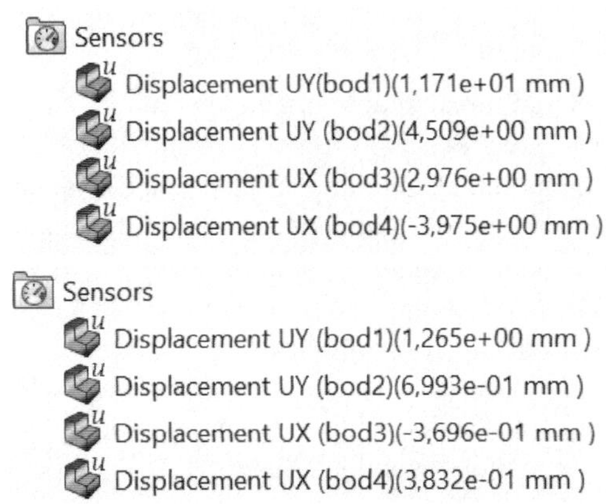


Fig. 10: Deformation of an Auxetic Material Under Compression (Lateral Contraction).

The displacements were used to calculate strain in both directions:

$$\varepsilon_x = UX_3 - UX_4, \qquad \varepsilon_y = UX_1 - UX_2, \tag{4), (5}$$

For the conventional model (Fig. 19), the computed values were:

$$\varepsilon_x = 6{,}951\,mm,\ \varepsilon_y = 7{,}201\,mm \ \rightarrow\ \nu = 0{,}965 \quad (6)$$

For the auxetic model (Fig. 20), the computed values were:

$$\varepsilon_x = -0{,}753\,mm,\ \varepsilon_y = 0{,}566\,mm \ \rightarrow\ \nu = -1{,}33 \quad (7)$$

These results confirm the auxetic behavior of the re-entrant model and are in good agreement with both the analytical and experimental values.

## 2.3 Experimental Validation

To validate the numerical and analytical results and to confirm the auxetic behavior of the re-entrant honeycomb structure, an experimental tensile test was performed. Both the conventional and re-entrant structures were modeled with dimensions of 11 × 10-unit cells and manufactured using 3D printing technology with PLA material (Fig11). To ensure accurate data acquisition, a measurement system (Fig. 12) was set up, consisting of: a tensile testing machine (1), a Dantec Dynamics high-speed camera (2), lighting (3), an adjustable tripod (4) and a computer system for image capture and processing (5). Precise positioning of the camera was essential to avoid optical distortions. The camera was therefore leveled to 0° tilt using spirit levels and calibration tools (Fig. 13).

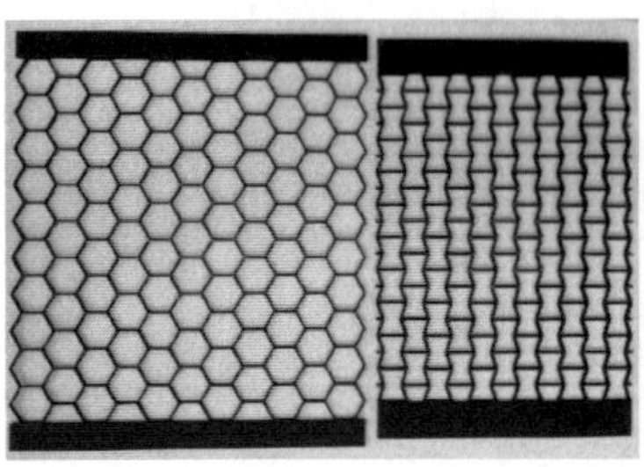

Fig. 11: 3D-printed PLA Samples.

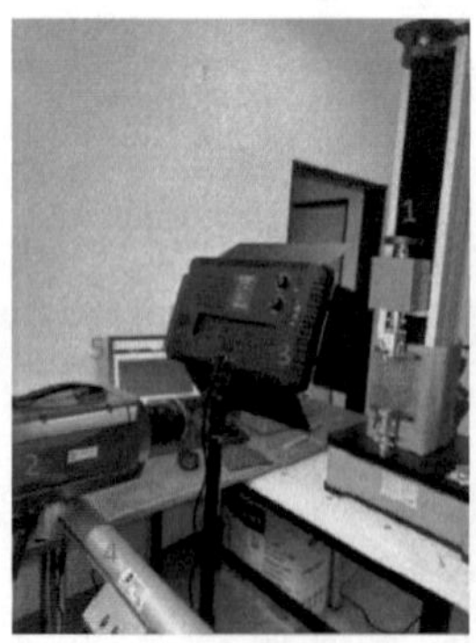

Fig. 12: Experimental Setup.

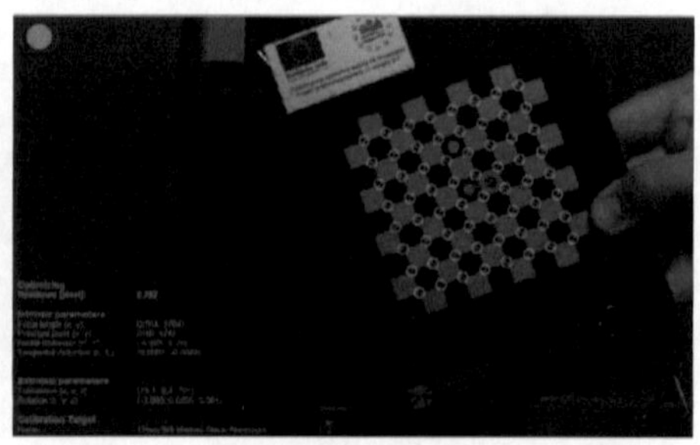

Fig. 13: Calibration of the Dantec Dynamics Camera.

To track deformation during the experiment, white reference markers were applied to the surface of both printed structures. The analysis was performed on the central region of the samples to minimize the influence of edge effects and clamping. The same four points used in the numerical simulation—points 1 and 2 aligned horizontally and points 3 and 4 on a second horizontal line—were selected for displacement analysis. These served to evaluate axial and transverse strains under tensile loading. Video recordings were processed using a custom MATLAB script, which identified the white points in images taken before and after deformation (Fig.14, 15) calculated the displacement vectors between the reference points, and computed the corresponding strain components. This approach enabled the calculation of Poisson's ratio directly from experimental data.

Poisson's ratio was calculated using (3), (4), (5):

- For the conventional structure, the computed value was:

$$\nu = -\frac{0{,}24}{-0{,}289} = 1{,}205 \quad (8)$$

- For the auxetic structure, the computed value was:

$$\nu = -\frac{0{,}043}{0{,}0591} = -1{,}375 \quad (9)$$

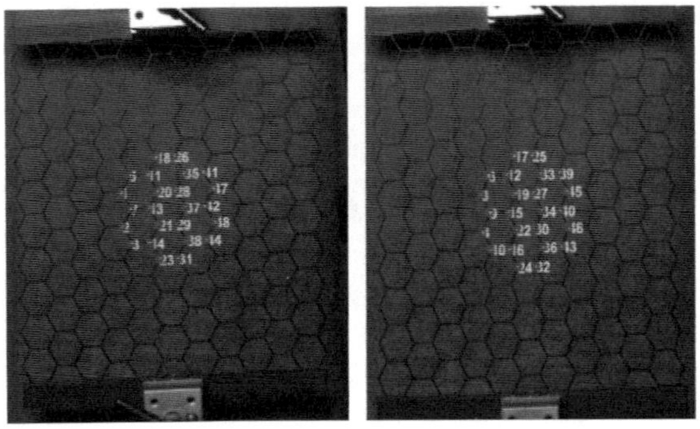

Fig. 14: Conventional Structure: Tracked Reference Points Before (Left) and After (Right) Deformation.

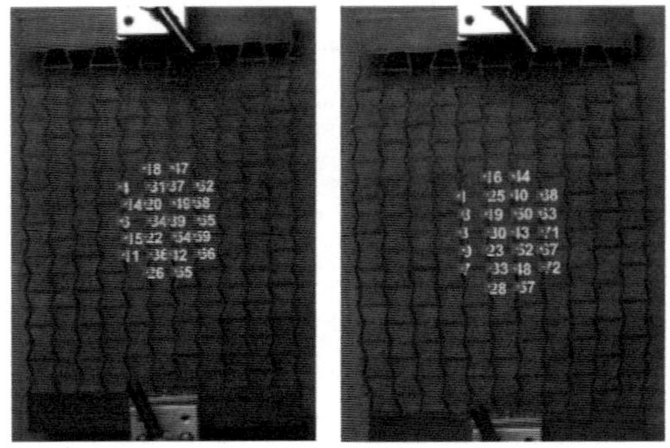

Fig. 15: Auxetic Structure: Tracked Reference Points Before (Left) and After (Right) Deformation.

Finally, Figure 16 illustrates the deformation of both structures before and after loading. Yellow outlines represent the undeformed geometry, while gray shows the deformed configuration, highlighting the inward contraction of the auxetic model.

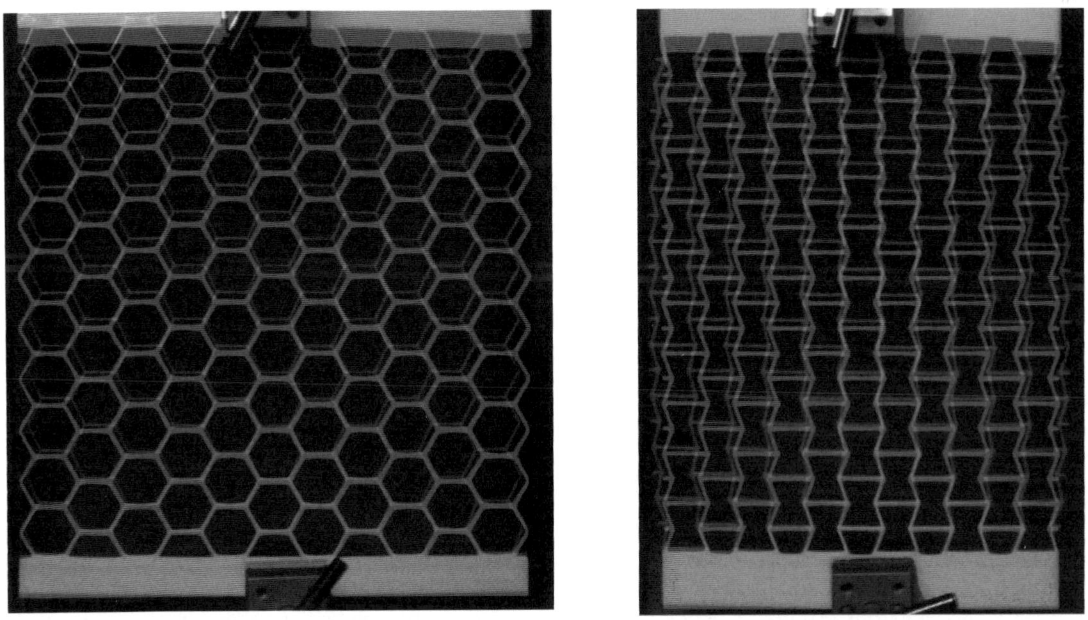

a) Conventional Structure

b) Auxetic Structure

Fig. 16: Comparison of Structural Deformation Before and After Loading.

## Conclusion

The results of this study confirm that auxetic structures exhibit fundamentally different mechanical behavior compared to conventional honeycomb geometries—primarily in their deformation response under axial loading and the resulting negative Poisson's ratio.

As shown in Table 1, Poisson's ratio for the conventional model remained positive across all evaluation methods, while the re-entrant auxetic structure consistently exhibited negative values, confirming its auxetic nature.

Tab. 1: Comparison of Poisson's ratio values for conventional and auxetic structures.

| Poisson´s ratio | Conventional | Auxetic |
|---|---|---|
| Analytically | 1 | -1,4 |
| Numerically (global) | 0,73 | -1,24 |
| Numerically (local) | 0,965 | -1,33 |
| Experimentally | 1,205 | -1,375 |

For the conventional honeycomb structure, the analytical, local numerical and experimental values show good agreement. However, the global numerical result (0,73) exhibits a significant deviation, suggesting that FEM simulations applied over the entire structure may underestimate transverse deformation or be sensitive to boundary conditions and mesh resolution.

In contrast, the auxetic re-entrant structure shows more consistent results across all methods. All values fall within a relatively narrow range, confirming the presence of auxetic behavior and the accuracy of the re-entrant design in achieving a negative Poisson's ratio.

These findings highlight the importance of combining multiple evaluation methods and considering local versus global deformation analysis. The results also demonstrate that auxetic structures, due to their higher geometric sensitivity and structural interdependence, may in some cases yield more stable numerical-experimental agreement than their conventional counterparts.

The discrepancies between analytical, numerical, and experimental values stem from several factors:

- Analytical models are based on idealized geometries and simplified assumptions, such as perfect elasticity and uniform boundary conditions. They do not account for nonlinear effects, material imperfections, or manufacturing tolerances.
- Numerical simulations, performed using the Finite Element Method (FEM), offer a more realistic approximation, capturing localized deformation, nonlinearity, and boundary conditions. However, they can still be influenced by factors such as mesh resolution, element type, or simplified material models.
- Experimental results represent the most realistic response but are inherently affected by manufacturing defects, clamping conditions, measurement precision, and camera calibration. Small shifts in 3D-printed geometry or minor asymmetries during clamping can significantly impact measured strains, especially for auxetic structures with high geometric sensitivity.

Despite these limitations, all methods independently validated the auxetic behavior of the re-entrant model and confirmed that topological design can successfully invert the sign of Poisson's ratio.

The findings contribute to a better understanding of auxetic structures and demonstrate that geometrical modifications alone can significantly influence macroscopic material behavior. This opens new possibilities for advanced engineering applications in:

- Impact protection and energy absorption,
- Flexible and conformal materials,
- Lightweight structural panels,
- Smart filters and biomedical devices.

This study serves as a baseline for future research, particularly in the experimental design of auxetic components and the optimization of geometry for targeted mechanical performance. Further work could explore: the influence of different unit cell parameters (e.g., wall thickness, internal angles), the use of alternative materials (including soft polymers or composites) and the behavior under dynamic or multi-axial loading conditions.

## Acknowledgement

The work was supported by the grant project KEGA 039TUKE-4/2024, VEGA 1/0516/22 and VEGA 1/0152/24.

## References

[1] Y. Liu, J. Zhang, P. Zhu, Auxetic meta-materials and their engineering applications: a review, Engineering Research Express 5 (2023) 42003, http://dx.doi.org/10.1088/2631-8695/ad0eb1.

[2] H. S. Gill, Mechanical and structure properties of cellular auxetic materials, Materials Today: Proceedings 37 (2021) 3320–3323.

[3] W. T. Nugroho, R. K. Cahyadi, A. B. Mulyono, Additive manufacturing of re-entrant structures: Well-tailored structures, unique properties, modelling approaches and real applications, Additive Manufacturing 78 (2023), https://doi.org/10.1016/j.addma.2023.103165.

[4] Y. Wang, L. Zhang, Y. Wang, Mechanical Properties of Re-Entrant Hybrid Honeycomb Structures for Morphing Wings, Biomimetics 9 (2024), https://doi.org/10.3390/biomimetics9010035.

[5] Q. C. Zhang, Z. H. Chen, L. J. Wu, Bio-Inspired Engineering of Honeycomb Structure - Using Nature to Inspire Human Innovation, Progress in Materials Science 74 (2015) 332–400.

[6] Y. Wu, X. Wang, C. Liu, Additively manufactured materials and structures: A state-of-the-art review on their mechanical characteristics and energy absorption, International Journal of Mechanical Sciences (2023) 108102.

[7] L. Yang, M. Harrysson, H. West, Mechanical properties of 3D re-entrant honeycomb auxetic structures realized via additive manufacturing, International Journal of Solids and Structures 75–76 (2015) 475–490.

[8] D. A. Yolcu, H. T. Erdogan, H. T. Gokozan, Measurement of Poisson's ratio of the auxetic structure, Measurement (2022) 112040.

[9] R. Gatt, L. Mizzi, J. Azzopardi, Hierarchical Auxetic Mechanical Metamaterials, Scientific Reports 5 (2015) 8395.

[10] Y. Yao, J. Lin, T. Zheng, A novel auxetic structure-based bone screw design: Tensile mechanical characterization and pullout fixation strength evaluation, Materials & Design 188 (2020) 264271.

[11] E. Bura, M. Bura, T. Kowalczyk, Theoretical and Applied Fracture Mechanics 138 (2025) 1–20.

[12] A. Alderson, An auxetic filter: A tuneable filter displaying enhanced size selectivity or defouling properties, Industrial & Engineering Chemistry Research 39 (2000) 654–665.

[13] X. Gong, et al., Impact Response of the Honeycomb Sandwich Structure with Different Poisson's Ratios, Materials (Basel) 15 (19) (2022) 6982.

[14] L. White, Auxetic foam set for use in smart filters and wound dressings, Urethanes Technology International 26 (4) (2009) 34–36.

[15] S. N. Orhan, Ş. Erden, Numerical investigation of the mechanical properties of 2D and 3D auxetic structures, Smart Materials and Structures 31 (6) (2022) 065011.

[16] M. Bera, Auxetic textiles, in: Functional and Technical Textiles, The Textile Institute Book Series, 2023, pp. 543–614.

[17] K. K. Dudek, et al., The multidirectional auxeticity and negative linear compressibility of a 3D mechanical metamaterial, Materials 13 (9) (2020) 2193.

# Identification of Paddle Motion During Kayak Paddling

B. Kropík[1,*], Z. Padovec[1], M. Fošnarič[2], T. Mareš[1]

[1] *Department of Mechanics, Biomechanics and Mechatronics, Faculty of Mechanical Engineering, Czech Technical University in Prague, Technická 4, 160 00, Prague, Czech Republic*

[2] *Department of Orthotics and Prosthetics, Faculty of Health Sciences, University of Ljubljana, Zdravstvena pot 5, 1000, Ljubljana, Slovenia*

* *bohumil.kropik@fs.cvut.cz*

**Abstract:** Kayak paddling on flat water was recorded using an Intel RealSense D455 depth camera mounted on the kayak, positioned in front of the athlete. The acquired data were analysed using Python libraries. The movement of the athlete's hands holding the paddle was tracked in RGB frames using the OpenCV CSRT tracker to facilitate the identification of motion relative to the camera coordinate system. The resulting output represents the movement of the paddle as a relative displacement within this coordinate system.

**Keywords:** kayak; paddle; motion analysis; motion tracking; depth camera

## 1 Introduction

Understanding the motion of both the kayaker and the paddle during sprint kayaking is crucial for enhancing performance. This knowledge not only supports training optimisation and equipment development but also contributes to scientific research in sports biomechanics. A review of existing literature on paddle movement was presented in [1], where paddle motion was categorised based on various positional types. For specific water-contact-defined positions, distinct phases and subphases were identified to facilitate training and research through an observational model.

Despite advancements, a practical setup for in situ measurement of paddle and kayaker motion has yet to be established. In [2], a motion tracking system using multiple cameras and markers was proposed to capture motion of a kayaker. A similar approach could be adapted to track the motion of the paddle shaft. However, modern depth cameras present a promising alternative, offering a favourable balance between complexity, accuracy, cost, and weight.

The system presented in this study was developed with the long-term objective of synchronising paddle motion data with previous research on paddle bending moments [3]. This integration aims to enable the estimation of the propulsive forces generated by the athlete during paddling.

## 2 Method

From a wide range of available depth cameras, the Intel RealSense D455 [4] stereo depth camera, shown in Fig. 1, was selected for this study. This camera uses stereo depth technology to estimate the distance of an object using two imagers and is also equipped with an RGB camera. A comparison was made between the main types of depth cameras: stereo depth cameras, Time of Flight (ToF) sensors, structured light sensors, and Light Detection and Ranging (LiDAR) sensors. Among these, the stereo depth camera emerged as the most suitable candidate.

LiDAR systems are typically more expensive and are designed to scan the surroundings, which can result in motion blur during rapid dynamic scenes. ToF and structured light cameras typically emit signals in the Near Infrared (NIR) spectrum. Since sunlight also contains NIR wavelengths, these systems are more susceptible to interference and increased noise during outdoor measurements. In contrast, stereo depth cameras often perform better in sunny outdoor conditions, provided that certain requirements are met, such as avoiding glare [5].

Fig. 1: Intel RealSense D455 depth camera used for data collection. [4]

Tab. 1: Selection of considered camera settings.

| Setting [-] | 1 | 2 | 3 |
|---|---|---|---|
| Resolution [px] | 1280x720 | 848x480 | 640x360 |
| FPS [Hz] | 30 | 60 | 90 |

Another factor in the decision was the performance at the expected measurement distance. The error in stereo depth systems is directly proportional to the square of the distance from the camera, whereas ToF systems benefit from longer signal return times. Therefore, while ToF cameras are more effective at longer distances, particularly near the upper limit of their measurable range, the short-range focus of this study favoured stereo depth technology. Other important decision criteria included resolution, optimal distance range, and frame rate. Based on a representative selection of identical settings for both the RGB and depth sensors in Tab. 1, this camera appeared to be a reasonable choice for the intended application. Considering that athletes can exceed 120 strokes per minute [6], the frame rate of the camera is crucial. The optimal operating range of 0.6 to 6 metres is well suited to measurement needs. Furthermore, the camera's ability to send a trigger signal for each frame was an important feature for future synchronisation with paddle measurement.

The camera was mounted on the kayak in front of the athlete, as shown in Fig. 2, at an approximate distance of 1.3 metres. A compact 3D-printed adapter and a GoPro holder were used to secure the camera in place, allowing it to capture both sides of the athlete, particularly both hands holding the paddle, using only a single camera. The holder was firmly attached to the kayak using silver tape. This non-invasive mounting method was chosen to avoid damaging the kayak and was feasible due to the low weight of the camera (116 grammes). The entire setup was designed to be as lightweight as possible so as not to interfere with the athlete.

Fig. 2: Schematic representation of a kayaker with the camera position indicated.

To control the camera and store data, Raspberry Pi 4 and Raspberry Pi 5 units powered by an external source were tested. However, due to the high demands on recording quality, these systems were not successful, as the settings listed in Tab. 1 resulted in frame drop issues. At lower frame rates, the Raspberry Pi 5 performed adequately, but the quality and quantity of data were prioritised over the weight of the setup. Therefore, a laptop was used for the final measurements to ensure reliable data acquisition. Paddling was recorded in a video file using Intel RealSense software [7] for subsequent evaluation and tracking. The camera configuration is a complex task, as numerous parameters affect the quality and density of the depth image. For this reason, the high-accuracy preset provided by the RealSense software was initially used.

Professional kayak paddles are typically feathered, and the angle between the blades usually remains constant during paddling. Considering the rotation of the paddle shaft throughout the stroke cycle, this provides an estimate of the boundary conditions. However, to fully track the rotational movement of the paddle, a system incorporating 3D printed balls mounted on a stick attached to the paddle was also tested. Initially, three red-painted balls, each with a diameter of 5 centimetres, were mounted on a 15-centimetre stick and attached to the paddle, as shown in Fig. 3. This configuration was intended to enable the description of both displacement and rotation.

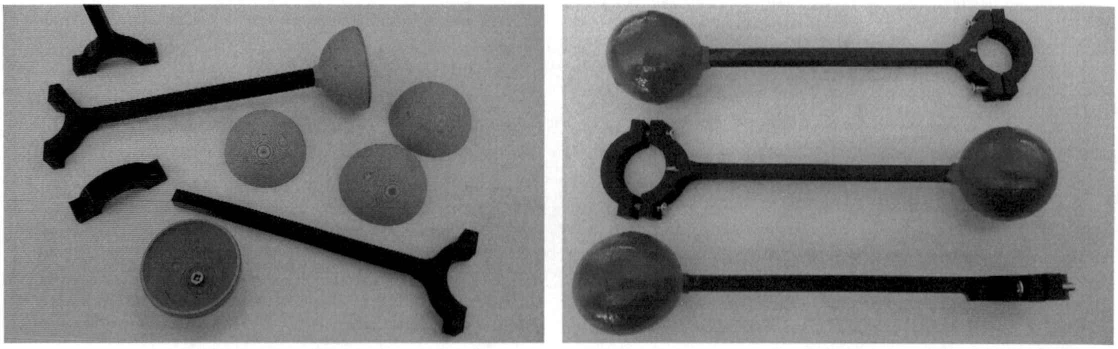

(a) Individual 3D-printed components (b) Assembled tracking solution

Fig. 3: 3D-printed setup for simultaneous displacement and rotation tracking.

However, when using only a single camera, it proved difficult to position the balls in such a way that they did not collide with the kayak, water or athlete, or overlap in the camera field of view during motion. Despite these challenges, the three-ball setup showed promise but would require at least two cameras to ensure that occluded sections in one recording could be supplemented by the other. As a result, the final measurement was conducted using only one ball, and hand tracking was considered instead for the initial concept test, as the hands differ in shape and colour from the paddle. Using this setup, approximately two minutes of data were recorded, including a slow start phase, regular paddling, and an end phase, with the primary focus on the regular paddling segment. A single recording of this duration can generate up to 20 gigabytes of data, highlighting the challenge of real-time data handling.

Since the camera measures the position of objects in the scene and represents them as a pixelated depth map, each pixel in the recorded data corresponds to a specific spatial coordinate relative to the camera's coordinate system. Therefore, it is only necessary to select suitable pixels that describe the movement of the paddle. Although manual selection would be possible, it is not practical given the large number of frames. For the initial paddling test, the same method was used that had previously proven effective in the author's earlier measurements involving short, planar motion, where the athlete's leg was analysed and points on the bars of a weight machine were successfully tracked. Following the same procedure, the recordings were evaluated using the Python wrapper [8] and OpenCV [9]. To track the hands and the ball, the discriminative correlation filter with channel and spatial reliability (CSRT) tracker was used, based on the method described in [10] and implemented in OpenCV. This tracker is generally capable of following an object during spatial movement and can dynamically adjust the size of the tracking box.

## 3 Results

To illustrate the nature of the recorded data, one RGB frame and the corresponding depth frame, visualised as a colour map, are presented in Fig. 4. As shown in Fig. 4 (b), the depth map does not contain a complete image; only those values that meet a certain precision threshold, based on the selected settings, are retained. Preliminary tests indicated that sunlight improves infill under the same settings. However, modifying the camera settings or applying filtering techniques can also enhance the infill, if necessary. Further insights on post-processing can be found in [11]. In this study, priority was given to quality over quantity, and the raw data were evaluated without any post-processing.

(a) RGB frame (b) Colourised depth frame

Fig. 4: Raw outputs from the depth camera.

The camera records raw RGB and depth frames, but for evaluation purposes, these must be spatially aligned. The alignment process may introduce slight distortions, as each pixel in the raw image is recalculated to match the coordinate system of the aligned image. Since it is possible to align the RGB image to the depth frame and vice versa, this capability results in four possible data configurations, potentially enhancing reliability and precision. An example of an RGB frame aligned to the depth frame is shown in Fig. 5. As most pixels in the depth frame do not contain valid depth information, the corresponding RGB pixels are excluded. The frames and results presented here were obtained using setting number 2, as specified in Table 1.

Fig. 5: RGB frame aligned with the corresponding depth frame.

Tracking fast and repetitive movements, such as paddling, proved to be challenging. Although the data obtained from the selected camera were of sufficient quality to describe the movement, which demonstrates the suitability of the camera for this application, the tracking approach itself was not effective. The chosen tracker frequently failed almost immediately after initiation. Various starting positions, bounding box sizes, and segments of the recording were tested during evaluation. While these parameters, along with other factors such as frame quality, movement speed, and background complexity, influenced the performance of the tracker, none of the tested configurations yielded satisfactory tracking results throughout the entire recording.

Fig. 6 presents a composite of selected segments from the RGB frames to illustrate the performance of the tracker. For clarity, the numbering of the boxes in the following description refers to their order from left to right in the figure. The first box shows the initial bounding box used to track the athlete's right hand, with a starting size of 51x51 pixels. The remaining boxes highlight different tracking behaviours. Even when the tracker did not fail entirely, its performance was often inadequate. For example, in the first box tracking the left hand, the tracker remained active but failed to follow the motion accurately. In the second box, which tracks the right hand, the resizing of the bounding box did not respond quickly enough to the movement.

Since the centre of the bounding box was initially used as a reference point for tracking, a complication arose when the box frequently shifted, causing its centre to no longer align with the paddle shaft, but instead

Fig. 6: Merged cut-outs showing hand tracking results across RGB frames.

rest on the edge of the hand, as illustrated by other boxes in the figure. This misalignment rendered the tracking results unreliable as they introduced distortion into the motion analysis.

To present the results of the tracking, Fig. 7 (a) shows the output from the beginning of the recorded data, identified using raw RGB frames with depth information aligned to them. The dots represent the position of the centre of the tracked bounding box in each frame over a period of almost five seconds, corresponding to 290 frames. This visualisation illustrates the tracked path independently, allowing it to be compared with the position of the hand in the original video frames. The tracker successfully captured one full cycle of the right hand (left line), but the tracking of the left hand failed almost immediately after initiation. Fortunately, since movement could be tracked in multiple frame types, the RGB frame aligned to the depth image provided a successful result for the left hand in the same section, as shown in Fig. 7 (b). Note the point located at the origin of the coordinate system, which typically indicates a complete failure of the tracker.

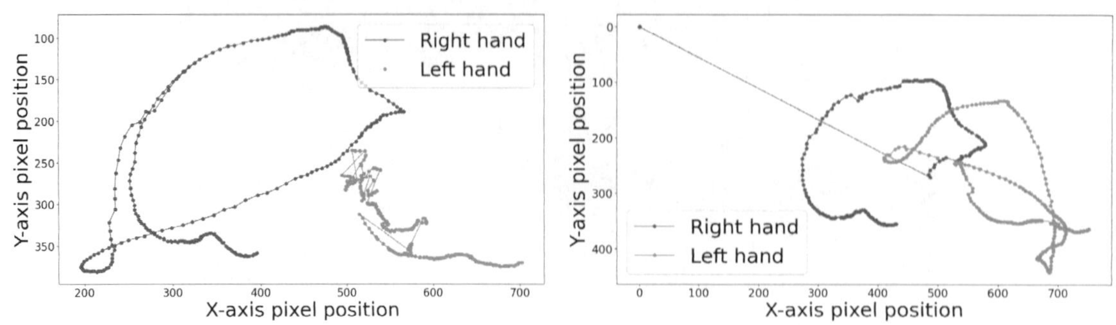


(a) raw RGB frame

(b) RGB frame alligned to depth

Fig. 7: Results of hand tracking in RGB image frames.

By merging the usable tracking results, as shown in Fig. 8 (a), a representative tracking output for one paddling cycle was obtained, where the bounding boxes followed the hands with reasonable accuracy. The evaluation of ball tracking was not included, as it was intended to provide supplementary information to the hand tracking. However, since the hand tracking did not yield sufficiently reliable results, it was not meaningful to continue with this additional analysis. Tracking in colourised depth images did not yield better results than those presented here, although it performed similarly and may be useful due to the reduced background information in such images.

Although the presented result is promising, it does not yet represent a reliable solution. The successful tracking segment is short as the tracker failed completely shortly afterwards. As discussed in the previous section, even this result is affected by distortion. On the other hand, once the pixels identified by tracking are known, the camera can directly provide their positions in three-dimensional space relative to its coordinate system. This is illustrated in Fig. 8 (b), which shows the result for the right hand in terms of distances (in metres) along the X, Y, and Z axes of the camera's coordinate system, presented as separate lines. The motion of the paddle can then be estimated as a change in these positions, assuming that the paddle shaft is approximately a straight line connecting both hands. A similar approach could also be applied to the setup with three tracking balls, where the spatial positions of the markers would allow for a more detailed reconstruction of both displacement and rotation of the paddle.

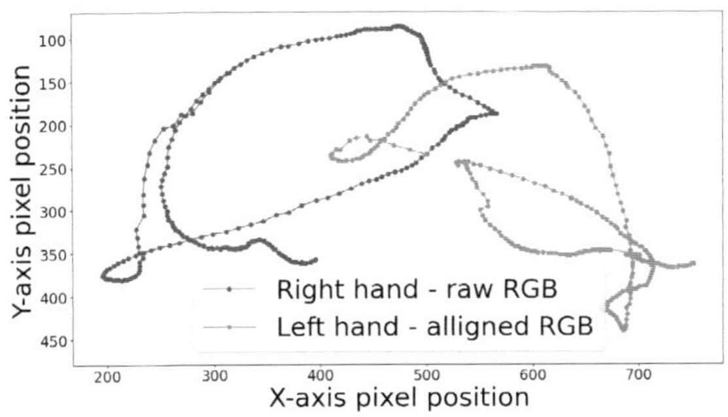

(a) Merged hand tracking data in pixel coordinates

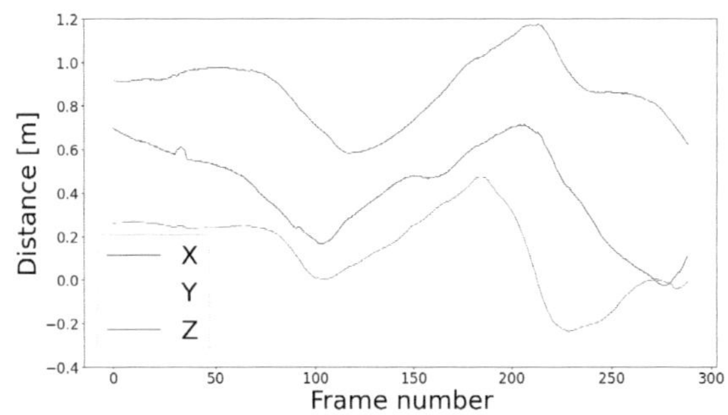

(b) Corresponding displacement of the right hand in the camera coordinate system along the X, Y, and Z axes

Fig. 8: Combined tracking results.

Due to the insufficient tracking output from the CSRT tracker, an RGB-based tracker [12] was briefly tested. This tracker followed the selected region accurately, but failed as soon as the region changed even slightly. In the context of this study, the hand holds the paddle, moves rapidly and rotates, resulting in limited and highly variable views of the hand. At certain points, the view is further disrupted by the paddle shaft. These conditions demonstrated that a different approach, such as a trained neural network for hand or ball recognition, would be more suitable to achieve reliable tracking throughout the entire recording. Even the ball, which is less affected by occlusion, failed under RGB tracking due to appearance changes. Considering the issue of overlapping and the dynamic nature of the scene, a trained tracker could offer significant advantages. Although the trackers used in this study generally perform well under standard conditions, they are not well suited to the specific demands of this application.

## 4 Conclusion

This pilot study demonstrated the feasibility of using a depth camera to track paddle motion during kayak paddling. The results confirm that the proposed approach is functional and holds promise for further development. In particular, it offers valuable potential when combined with previously conducted strain measurements [3], enabling the estimation of the propulsive force exerted on the paddle shaft during each stroke.

Beyond paddle tracking, the system may be extended to monitor specific body segments of the kayaker or to simultaneously track both the athlete and the paddle using multiple synchronised cameras, strategically positioned to minimise occlusion. While the depth camera itself performed reliably, the evaluation process, especially the use of a tracker for automated data analysis, requires further refinement. The large volume of data generated makes manual evaluation impractical.

Future work will focus on integrating more advanced tracking techniques, ideally leveraging models trained specifically to detect and track paddle motion and relevant body parts. This will enhance automation, accuracy, and overall utility of the system for performance analysis and biomechanical research.

## Acknowledgement

This study received support from the Grant Agency of the Czech Technical University in Prague, under grant No. SGS24/123/OHK2/3T/12.

# References


[1] L. K. McDonnell, P. A. Hume, V. Nolte, An observational model for biomechanical assessment of sprint kayaking technique, Sports Biomechanics 11:4 (2012) 507–523, doi: 10.1080/14763141.2012.724701.

[2] S. Miyazaki et al., Development of a video camera-type kayak motion capture system to measure water kayaking, PeerJ 11:e15227 (2023), doi: 10.7717/peerj.15227.

[3] B. Kropík, K. Doubrava, T. Mareš, Identification of the Bending Moment on the Paddle of a Kayaker, in proc.: EAN 2023 - 61st Conference on Experimental Stress Analysis, Proceedings of Full Papers (2023), ed. M. Hagara, R. Huňady, Technical University of Košice, Košice, 130–135.

[4] Retrieved from: https://www.intelrealsense.com/depth-camera-d455/

[5] Retrieved from: https://dev.intelrealsense.com/docs/tuning-depth-cameras-for-best-performance

[6] B. B. Gomes et al., Paddling time parameters and paddling efficiency with the increase in stroke rate in kayaking, Sports Biomechanics 21:10 (2022) 1303–1311, doi: 10.1080/14763141.2020.1789204.

[7] Retrieved from: https://www.intelrealsense.com/sdk-2/

[8] Retrieved from: https://github.com/IntelRealSense/librealsense

[9] Retrieved from: https://docs.opencv.org/4.x/d6/d00/tutorial_py_root.html

[10] A. Lukežič et al., Discriminative correlation filter tracker with channel and spatial reliability. International Journal of Computer Vision 126 (2018) 671-–688. doi: 10.1007/s11263-017-1061-3

[11] Retrieved from: https://dev.intelrealsense.com/docs/depth-post-processing

[12] Retrieved from: https://opensourcephysics.github.io/tracker-website/

# On a Possibility of High-Resolution Neutron Diffraction Studies of High-Entropy Ceramics Substructure

P. Mikula [1,a], V. Ryukhtin [1], J. Dusza [2],

[1]Nuclear Physics Institute CAS, 250 68 Řež, Czech Republic

[2]Institute of Materials Research of SAS, 040 01 Košice, Slovak Republic

[a]mikula@ujf.cas.cz

**Abstract:** In our contribution we present the methodology results of the feasibility of using high-resolution neutron scattering for studies of some additional substructure parameters of the samples of H-E (High-Entropy) ceramics in relation to their technological preparation. For demonstration we used 4 different samples, namely, 2 monocarbides HfC and TAC and 2 H-E carbides HEC1-(Hf-Ta-Zr-Nb)C and HEC2-(Mo,Nb,Ta,V,W)C. The neutron diffraction results could be used in some special studies exploiting diffraction line profile analysis for e.g. micro-strain/micro-stress analysis of plastically deformed samples, determination of possible phases having very close values of the lattice spacings. On the other hand, the small-angle scattering provides the results e.g. mean radius of pores, size distribution of inhomogeneities in the homogeneous matrix. The samples were of the half-moon disc of the diameter of $\phi = 15$ mm and the thickness from 1.1 mm to 1.7 mm.

**Keywords:** neutron diffraction; small-angle-neutron scattering; high-entropy-ceramics; substructure.

## 1 Introduction

H-E alloys and ceramics have superior physical and mechanical properties (e.g. exceptional high-temperature strength/hardness), which naturally depend on their structure (as studied by neutron and/or X-ray diffraction) and microstructure (as studied e.g. by scanning electron microscopy (SEM), transmission electron microscopy (TEM) small-angle neutron scattering (SANS) and others) [1, 2]. H-E samples typically consist of five or more elements that are mixed in equimolar concentrations. In this case we used for the studies the neutron diffraction and SANS instruments which are at disposal at the reactor in Řež (see https://www.ujf.cas.cz/cs/oddeleni/oddeleni-neutronove-fyziky/ and (https://cz.linkedin.com/company/research-centre-rez).

## 2 Experimental details

### 2.1 High-resolution neutron diffraction

The high-resolution 3-axis setting as schematically shown in Fig. 1 is in fact powder diffractometer optimized for a chosen reflection. Its best resolution can be achieved when optimizing the diffractometer parameters and thus minimizing the dispersion of the whole system [3].

First, we looked at the basic structure by conventional powder diffractometer where angular position of the diffraction maxima provides information on lattice changes related to individual H-E samples (see Fig. 2).

The structure of all samples is single phase Fm-3m. However, the lattice parameter of the individual samples is slightly different which results in the mutual $2\theta$–shift of the peaks related to measured reflections. In further measurements we focused on the reflection marked by arrow in Fig. 2. The related result is shown in Fig. 3. The different level of substructure quality of the individual samples is evident. The best quality material can be seen in the case of TAC sample, where the corresponding *FWHM* (Full Width at Half Maximum) can be considered as a minimum value corresponding to the standard sample [3]. The broadening of the other rocking curves results from the porosity, grain size and plastic deformation arising from the sample preparation.

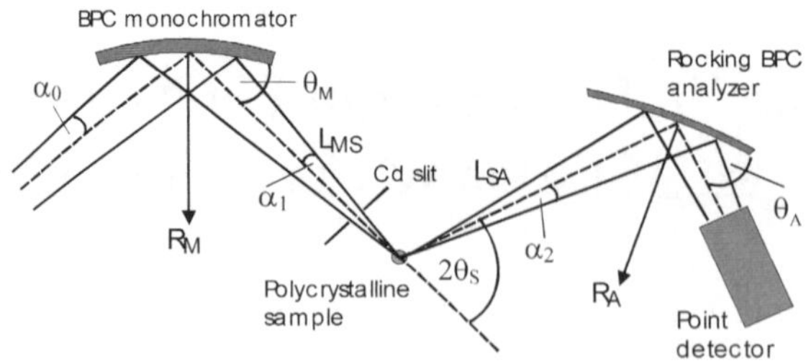


Fig. 1: Three-axis diffractometer setting employing BPC monochromator and analyzer as used in the experiment ($R_M$, $R_A$ - radii of curvature, $\theta_M$, $\theta_A$ - Bragg angles, $L_{MS}$, $L_{SA}$ - distances). Diffraction profiles are obtained by rocking the BPC analyzer.

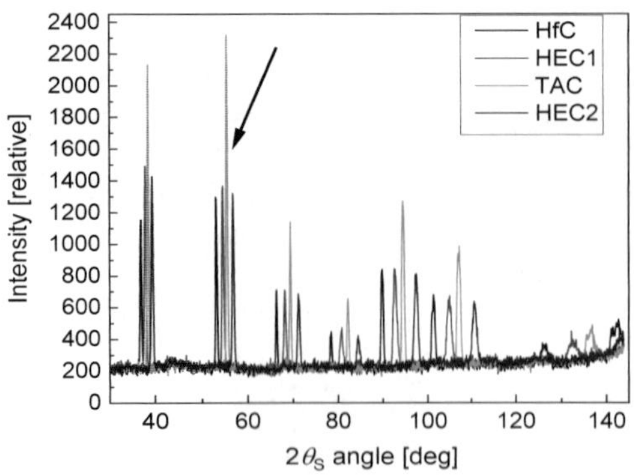


Fig. 2: Conventional powder diffraction spectra related to the individual H-E samples.

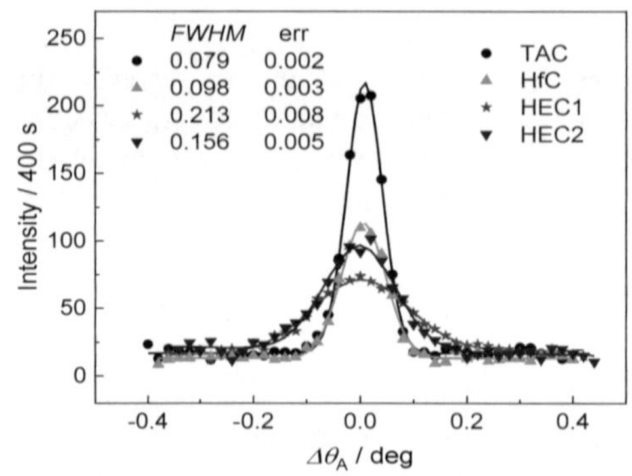


Fig. 3: Diffraction profiles obtained by rocking the analyzer in the vicinity of the Bragg reflection marked by arrow in Fig. 2.

## 2.2 High-resolution neutron small-angle neutron scattering (SANS)

Less known SANS method is generally used for investigation of rather large objects of the average dimension from 1 nm to 10 μm: Complex fluids, alloys, precipitates, biological assemblies, ceramics, glasses, flux lattices, long-wavelength CDWs and SDWs, critical scattering, porous media, fractal structures etc. Bulk samples can be used (1 or 2 mm thick). SANS is sensitive to the light elements (H, C and N) as well as isotopes such as H and D. Overview of applications related to usually used methods is in the following scheme:

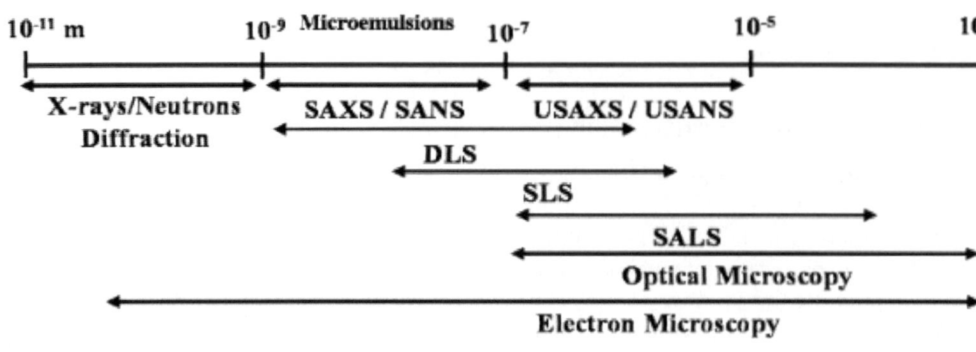


Fig 4. Overview of the useful characterization methodologies.

Usually, 3 types of SANS instruments operating with different resolution and the range of the momentum transfer $Q$ can be found in world neutron laboratories:

| | **Size range ($R$) of inhomogeneities** |
|---|---|
| Pin-hole (conventional) SANS instrument | 1 nm … 100 nm |
| DBC diffractometer with bent crystals [4] | 10 nm … 1 µm |
| Ultra DC-SANS setting with perfect crystals | 1 µm … 10 µm |

Each of them operates in the adequate range of the momentum transfer $Q$ as shown in the next Fig. 5.

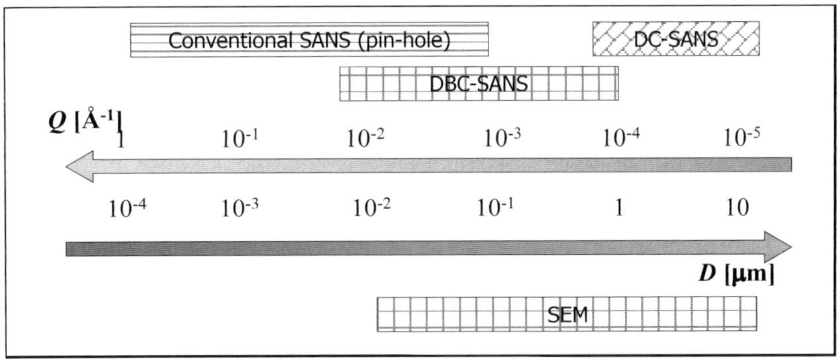


Fig. 5: $Q$ or $R$ range comparison of the individual instruments.

In our case the double-crystal SANS diffractometer operating with neutron wavelength of 2.09 Å is designed for the measurements of small-angle neutron scattering with a high momentum-transfer $Q$-resolution. The fully asymmetric diffraction geometry of the bent Si analyzer is employed to transfer the angular distribution of the scattered neutrons to the spatial distribution and to analyze the scattering curve by the 2-d PSD (position sensitive detector). The remote control of the curvatures of the monochromator and analyzer crystals makes it possible to tune the instrument resolution in the $Q$ range from $10^{-4}$ to $10^{-3}$ Å$^{-1}$, according to the studies of expected size of structural or compositional inhomogeneities (see Figs. 5 and 6). It is mainly used for investigations of the inhomogeneities (e.g. in alloys, porous media), cavities, fractal structures, precipitates

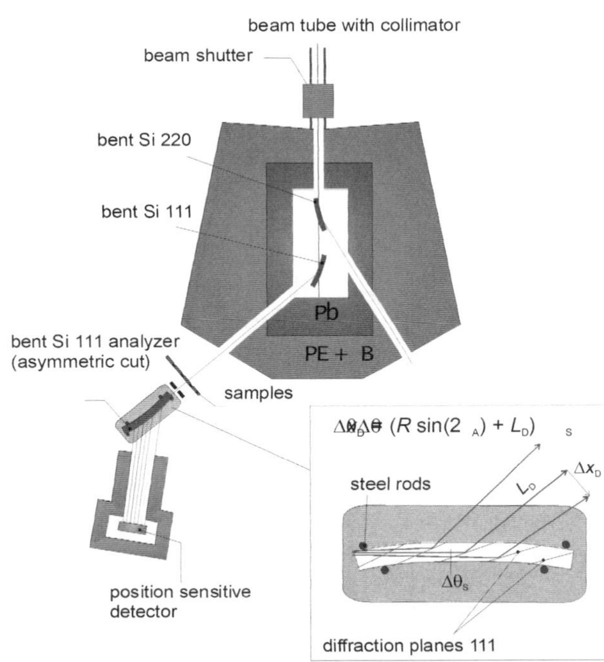


Fig. 6: Schematic diagram of the DBC-SANS diffractometer. Inserted part shows the transformation of the scattering angle deviation on the position coordinate.

etc. in the size range from about 50 nm to several μm. Fig. 7 shows an example of the obtained SANS curves, Fig. 8 obtained fitted volume weighted size distributions and Tab.1 summarized the results.

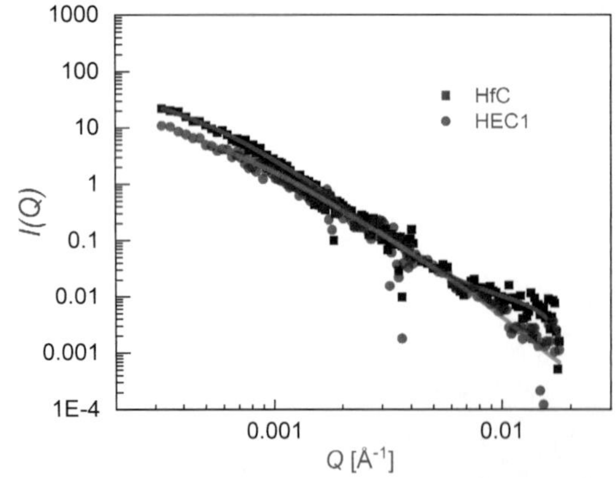


Fig. 7: SANS curves of the HFC and HEC1 samples with fitted lines.

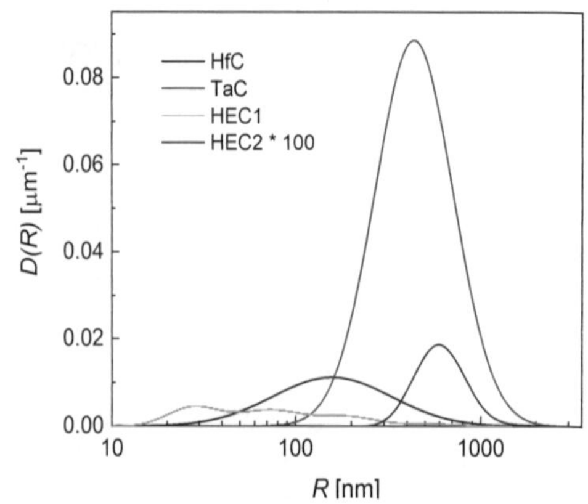


Fig. 8: Fitted volume weighted size distributions.

Tab.1: Porosity parameters obtained from the of the data fitting.

| | HfC | TAC | HEC1 | HEC2 |
|---|---|---|---|---|
| **Volume fraction** | 0.425 ± 0.019 % | 5.296 ± 0.03 % | 0.103 ± 0.011 % | 0.009 ± 0.014 % |
| **Mean radius of pores** | 355 ± 28 nm | 621 ± 3 nm | 261 ± 35 nm | 720 ± 71 nm |
| **Model used** | LogNormal distribution (w=0.49) of spherical pores | LogNormal distribution (w=0.49) of spherical pores | Spherical particles with size distribution of 7 splines | Spherical particles with size distribution of 7 splines |

# 3 Conclusion

Feasibility of the high-resolution three-axis neutron powder diffractometer and small-angle scattering diffractometer were tested for a possible additional study of structure and substructure of H-E samples. The measurements were carried out on four H-E samples and the analyzer rocking curves (sensitive on plastic deformation and/or mean grain or inhomogeneity size [5, 6]) are shown in Fig. 3 where different structure quality is manifested on the related *FWHMs*. However, a closer inspection of Fig. 3 reveals the detail in the asymmetry of the two peaks HEC1 and HEC2, which are skewed towards lower $\Delta\theta_A$ values. This should be related to the incomplete diffusion of some monocarbides (ZrC and potentially the HfC too in the case of HEC1) as has been reported in paper [2]. The porosity of the samples was measured by the small-angle neutron scattering and the results of the porosity parameters and volume weighted size distributions obtained from the evaluation of the SANS data are presented in Fig. 8 and Tab. 1. It can be stated that both presented high-resolution neutron-scattering methods can offer an additional useful average structural information of bulk samples and thus offer support to complement the information achieved by using the other conventional characterization methodologies (e.g. with electron microscopy techniques). We believe that it can be interesting for potential external users who are dealing with the structure and substructure parameters of the H-E alloys and ceramics and would like to apply for a beam time by means of the CANAM project at some of the instruments [7].

# Acknowledgement

Measurements were carried out at the CANAM instrument of NPI CAS Řež installed at the CICRR infrastructure, which is financially supported by the Ministry of Education and Culture - project LM2023041. The authors acknowledge support from ESS participation of the Czech Republic – OP

(CZ.02.1.01/0.0/0.0/16_013/0001794) and from the project ESS Scandinavia-CZ II (LM2018111), respectively. Furthermore, they acknowledge support from the CAS in the frame of the program "Strategie AV21, No. 23". The authors thank B. Michalcová from NPI CAS for significant help with measurements and data elaborations.

## References

[1] M.-H. Tsai, J.-W. Yeh, High-Entropy alloys: a critical review, Mater. Res. Lett. 2 (2014) 107-123. https://doi.org/10.1080/21663831.2014.912690.

[2] E. Castle, et al., Processing and propertied of high-entropy ultra-high temperature carbides, Scientific Reports 8 (2018) 8609-8621. DOI:10.1038/s41598-018-26827-1.

[3] P. Mikula, V. Ryukhtin, P. Strunz, On the resolution of the three-axis neutron diffractometer setting optimized for some powder diffractometry studies, Journal of Neutron Research, IOP Press 26, Issue 1 (2024) 15-33. DOI:10.3233/JNR-240001.

[4] J. Šaroun, et al., Focusing doubly-bent-crystal diffractometer in combination with PSD for SANS experiments, Journal de Physique 3 C7 (1993) 439-442. DOI:10.1051/jp4:1993890.

[5] R. Delhez, T.H. De Keijser, E.J. Mittemeijer, Determination of crystallite size and lattice distortions through X-ray diffraction line profile analysis, Fresenius' Zeitschrift für analytische Chemie 312 (1982) 1-16. https://doi.org/10.1007/BF00482725.

[6] V. Davydov, et al., Single-line diffraction profile analysis method used for evaluation of microstructural parameters in the plain ferritic steel upon tensile straining, Materials Science Forum 571-572 (2008) 181-188. https://doi.org/10.4028/www.scientific.net/MSF.571-572.181.

[7] https://www.ujf.cas.cz/cs/vyzkum-a-vyvoj/open-access/.

# Residual Stress Measurement in the Cold Spray Layers

Radka Musilová[1], Marek Vostřák[1], Andrij Krasij[1], Jaroslav Václavík[1] and Michal Švantner[2]

[1] *Dynamic Testing Laboratory, Research and Testing Institute Pilsen, Tylova 1581/46, 301 00 Plzeň, Czech Republic, musilova@vzuplzen.cz*

[2] *New technologies Research Centre University of West Bohemia in Pilsen, Univerzitní 8, 301 00 Plzeň, Czech Republic*

**Abstract:** The residual stresses in cold spray layers were measured on several samples in different layer condition. The common hole-drilling method [1] was used for residual stress investigation.

**Keywords:** residual stresses, hole-drilling method, cold spray.

## 1 Introduction

Coatings deposited by using the cold spray method are used in industry sectors such as aerospace, electronics, energy, automotive, etc. In the following experiment, specimens with different heat treatment will evaluate this influence on residual stresses [2].

Cold spray technology for coating deposition used the high kinetic energy of particles impacting the substrate surface. This intense kinetic energy enables the deposition of material particles to form coatings that are free from oxide inclusions, exhibiting high density and excellent cohesive strength [3].

Due to the relatively low process temperature which is maintained below the melting point of the powder is not created heat-affected zone, and the original microstructure of the powder remains the same. Furthermore, the process is characterized by high deposition efficiency, chart of technology see Fig. 1.

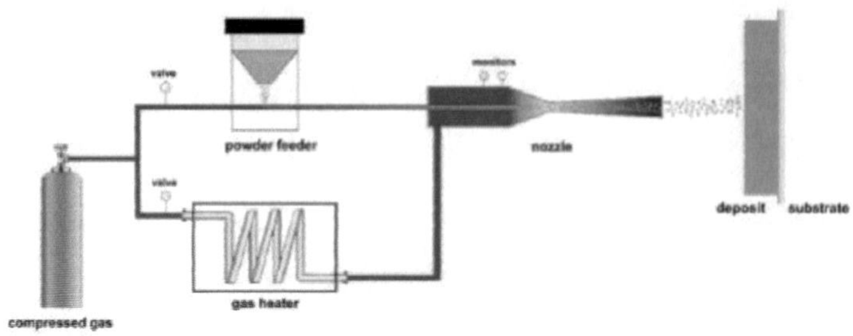


Fig. 1: Chart of deposition process [2].

### 1.1 Used materials

The development of cold spray process was focused on optimizing the technological parameters of coating layer. As a substrate was used common steel ČSN 11 353, which is used for pipes industry. The microstructure of this steel is consistent with ferrite and perlite, see Fig. 2 a). For cold spray was used stainless steel material specifically dust Praxair FE 101 (316L), with dimension 45 μm of particles, the picture of the dust it is possible to see on Fig. 2 b). This type of material has very good corrosion properties, and the layers are suitable for protection against abrasion, cavitation and erosion.

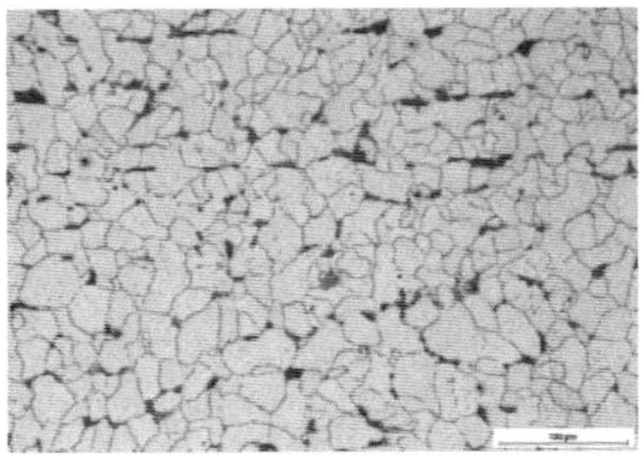

a)

b)

Fig. 2: Microstructure of the used materials: a) material for substrate; b) material for cold spray.

## 1.2 Coating parameters and heat treatment

The spraying was carried out using the Impact Spray Sytem, Gun 6/112 EvoCSII. Thickness of the layer was 1.8 mm. Main goal was introduce compressive stress into the surface layer.

Tested samples were prepared with the same coating parameters, corresponding to the main optimalization, see Tab. 1.

**Table 1.** Deposition parameters.

| **Layers param.** | | | | |
|---|---|---|---|---|
| **t [°C]** | **Press [bar]** | **Distance [mm]** | **No. of cross** | **ϕ [m³/h]** |
| 1100 | 50.0 | 30.0 | 14 | 3.0 |

The heat treatment of the coating after spraying was carried out to determine its effect on the final microstructure of the coating, particularly regarding as possible decreasing the porosity and better mechanical properties. For each temperature was used only one sample, see Table 2. It was prepared and measured sample without heat treatment as a reference sample, marked as "A".

**Table 2.** Heat treatment parameters.

| **Temperature T [°C]** | **Time t [min]** | **Coolant** | **Marking** |
|---|---|---|---|
| Without | - | - | A |
| 750 | 60 | Air | B |
| 900 | 60 | Air | C |
| 1050 | 60 | Air | D |

# 2 Residual stress measurement

Sample dimensions for residual stress measurement was 25 x 30 x 5 mm, for more detail see Fig. 3. The arrow on the sample shows the direction of applying the coatings.

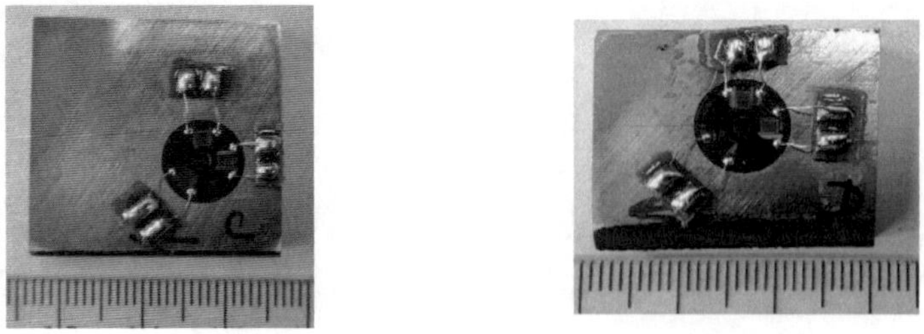

Fig. 3: Specimens for residual stress investigation.

Residual stress measurement was realized using MTS3000 – Restan, see Fig. 4 and evaluated using software EVAL 8.01 (05-2024), both HBM SINT producer. The methodology for calculating was used ASTM E837 ADV Not Uniform.

a) b)

Fig. 4: MTS3000 Restan system (a) with tested specimen (b).

# 3 Results

Evaluated results are presented in Fig. 5 for all tested samples. It is evident that the samples under the heat treatment have the same trend of measured residual stresses. On the surface compression state turns nearly from the depth of 0.1 mm to tension values. On the other hand, in the sample without heat treatment only tension values were observed.

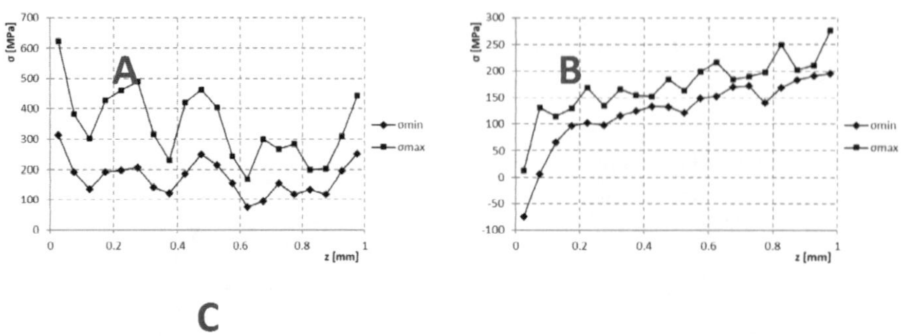

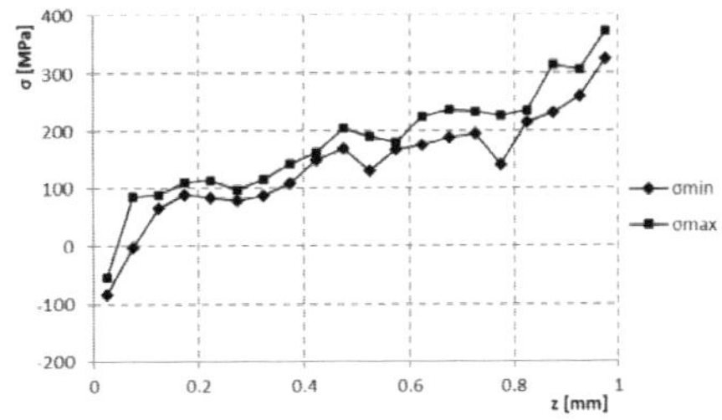

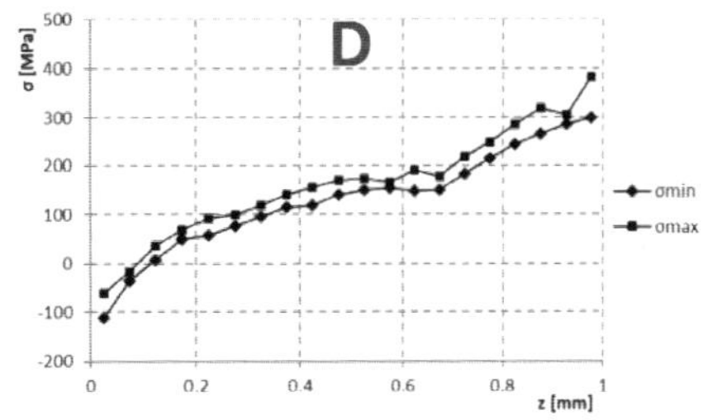

Fig. 5: Resulted residual stresses in specimens A to D.

On Fig. 6 and Tab. 3 the residual stress is depicted in 1.0 mm depth. Specimen B shows lower values than other specimens. This may be caused by lower temperature heat treatment, due to this occurred so-called "recovery", which leads precisely to a reduction the residual stress. On the contrary, for specimens C a D, where the temperature of heat treatment was higher, the material already completely recrystallized. Here, nucleation of new grain without signs of deformation, these grains are more stable that the surrounding "deforms" ones. The result is therefore a new structure whose properties are closer to those of the undeformed material. Due to this, these samples had almost the same residual stress value as the sample without heat treatment.

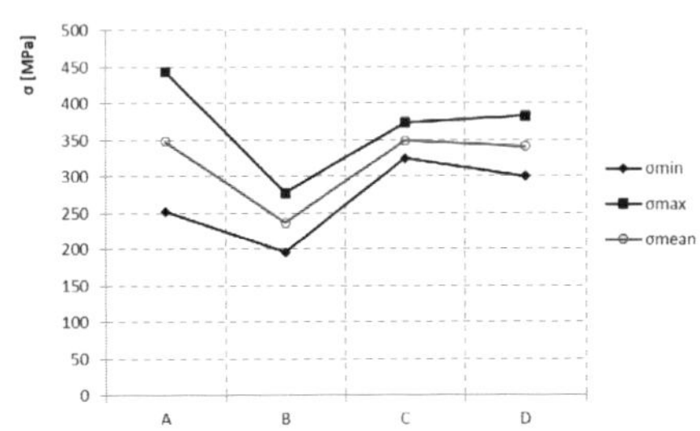

Fig. 6: Results of residual stress measurements in 1.0 mm depth.

A different trend can be observed in the distribution of residual stress across the measured depth of the specimen. Heat treated specimens on the surface exhibit compressive values, which continuously transition into tensile ones. However, the specimen without heat treatment shows tension values. This phenomenon is likely caused by the layer process itself, where the individual layers are deposited under a pressure of 50 bar at high temperature, where the resulting in continuous deformation of the deposited material ("forming"). Subsequent heat treatment (after such intense "forming") removes only the tension values from the surface. Additionally, precipitation of secondary phases may also occur.

**Table 3.** Residual stress on the 1.0 mm depth

| | **Without HT** | **750 °C** | **900°C** | **1050°C** |
|---|---|---|---|---|
| | A | B | C | D |
| $\sigma_{min}$ | 252 | 195 | 324 | 299 |
| $\sigma_{max}$ | 443 | 277 | 373 | 382 |
| $\sigma_{mean}$ | 347 | 236 | 348 | 340 |

## 4 Conclusion

Based on these measurements, we can conclude that heat treatment has a significant impact on the change in stress on the surface of the specimen. Compressive stress on the surface of the specimen is desirable. However, it is important to mention that subsequent measurements are necessary to determine if the depth of compressive stress achieved is sufficient for subsequent use in the pipeline industry.

## References


[1] ASTM Determining residual stresses by the hole-drilling strain-gage method. Standard test method E837-20. American Society for Testing and Materials, West Conshohocken, PA (2013).

[2] Dlouhá, Ž., Duliškovič, J., Vostřák, M., Vliv procesních parametrů a tepelného zpracování na 316L povlaky aplikované technologií Cold Spray, VZU-VZ-37/24/021 Research report, Research and Testing Institute Pilsen, (2025)

[3] Gabani, D., Gibmeier, J., Residual stress in cold gas sprayed Titanium coatings – role of substrate material and process parameters, Institute for Applied Materials, Proceedings of 11th European Conference on Residual Stresses ECRS 11, Prag, Czechia, July 2024. ISBN 1211 5894 (print)

# Validation of Polymer Material Model Using Tensile Test Simulations and Mesh Sensitivity Analysis

P. Palička [1,a], M. Hagara [1], B. Schürger [1], R. Huňady [1,2]

[1] *Technical University of Košice: Letná 1/9, 042 00 Košice, Slovakia; peter.palicka@tuke.sk, martin.hagara@tuke.sk, barbara.schurger@tuke.sk, robert.hunady@tuke.sk*
[2] *VSB- Technical University of Ostrava: 17. listopadu 2172/15, 708 00 Ostrava-Poruba, Czech Republic; robert.hunady@vsb.cz*
[a] *peter.palicka@tuke.sk*

**Abstract:** In crash simulations, precise material modeling is essential to ensure the accuracy of numerical results. Polymer materials, due to their complex deformation behavior under dynamic loading, require carefully validated models. This work focuses on validating material card developed for ASA polymer, which are commonly used in the automotive industry. The objective was to verify the accuracy of material card by comparing simulation results with experimental tensile data across three strain rates $10^{-2}$, $10^{0}$ and $10^{2}$ $s^{-1}$. In addition, a mesh sensitivity analysis was performed to assess the effect of element size on simulation outcomes. The findings confirm the reliability of the developed models and provide recommendations for their application in crash simulations. Finite element simulations were performed in the VPS solver. The simulations aimed to replicate experimental tensile tests at three strain rates: $10^{-2}$, $10^{0}$ and $10^{2}$ $s^{-1}$. Stress-strain relationships and failure curves (maximum strain vs. strain rate) were obtained experimentally using tensile, flexural and DMA tests and used to define the material behavior. A shell model was used with defined boundary conditions to ensure simulation accuracy. Tensile force and elongation were recorded and compared with experimental results. For the highest strain rate, when elongation could not be measured, force-time curves were used instead.

**Keywords:** polymer; ASA; material model; FEM; failure.

## 1 Introduction

In general, the basic mechanical properties of materials, such as modulus of elasticity, Poisson's ratio, yield strength or ultimate strength, which are normally given in the material data sheet, are obtained from standard tests conducted under quasi-static loading conditions and at a temperature of 20 °C. In the case of dynamic conditions, where the response of the material depends significantly on the loading rate or temperature, a more comprehensive description of its behavior is required. The determination of the mechanical properties of plastics has received considerable research attention in practice, as evidenced by the large number of technical publications and scientific articles. A detailed view of the subject is provided by the authors in a review paper [1], in which they discuss methods of dynamic testing of plastics, examining the mechanical properties of several classes of polymers, not only as a function of strain rate and temperature, but also as a function of pressure. The work [2] provides a complex review of the methods and equipment used in high-speed testing, ranging from drop-type machines, to split Hopkinson rods, to ballistic testing equipment. At the same time, the work gives a basic description of optical methods suitable for dynamic displacement and strain measurements. These are mainly methods based on the principles of interferometry, photogrammetry or thermography.

In papers [3-5], the authors analyzed the influence of strain rate and temperature on the mechanical response of different polymers. The strain rate was considered in the range from $10^{0}$ $s^{-1}$ up to $10^{2}$ $s^{-1}$ and the temperature from -40 °C up to glass transition temperature $T_g$. The presented results of their research confirm the significant influence of the above factors on the mechanical properties of the tested materials. In general, at higher temperatures, plastics behave more ductile, their ultimate strength decreases and their ductility increases. A similar effect on the mechanical properties of plastics is also observed as the loading rate decreases.

Conversely, at lower temperatures, plastics behave more brittle, with an increase in ultimate strength, while the overall strain decreases, similar to the increase in loading rate.

## 2 Experimental Testing and Results

The raw material used for the production of the test specimens was ASA polymer granulate. Plastic sheets with dimensions of 525 x 168 mm and a thickness of 2.5 mm were produced first, by injection moulding. The investigated material was tested in the direction of melt flow into the mould (0°). The test specimens were waterjet cut from the plates according to the specified vector trajectory. The specimens for tensile tests were of a standard shape with a narrowed cross-section in the area of evaluation (Fig. 1a). The specimens for bending and DMA tests had a rectangular shape (Fig. 1b).

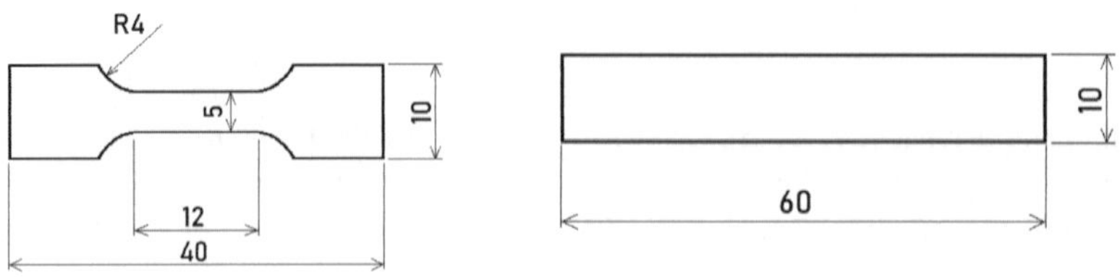


a) Tensile test specimen b) Flexural and DMA test specimen

Fig. 1: Test specimens.

Tensile tests were conducted to describe the mechanical behavior of the materials under tensile stress, and tests were conducted until failure of the specimens. Seven specimens were tested at three strain rates $10^{-2}$ $s^{-1}$, $10^{0}$ $s^{-1}$ and $10^{2}$ $s^{-1}$ and temperature 23 °C. Tests at the first two rates were conducted on a LabTest 5.050ST universal tensile machine. The mechanical properties at the highest rate were tested on a machine developed by company Lenam, s.r.o.

**Tensile tests at strain rates $10^{-2}$ $s^{-1}$ and $10^{0}$ $s^{-1}$**

The test device and the method of fixing the specimen are evident from Fig. 2. The specimen was fixed with pins whose radius corresponded to the curvature of the specimens.

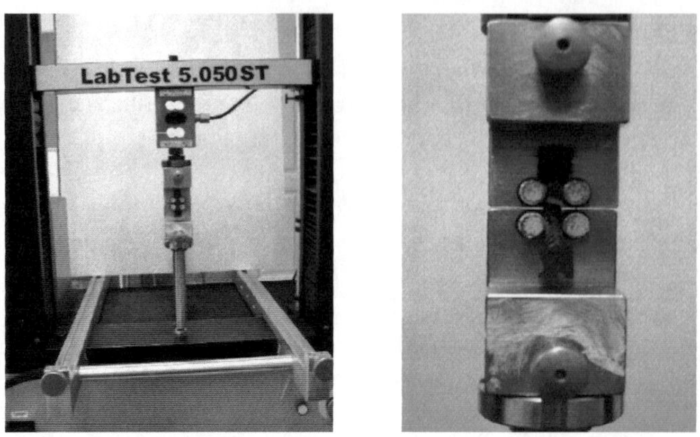


a) LabTest 5.050ST machine b) Tensile test

Fig. 2: Tensile test on universal machine LabTest 5.050ST.

The tensile force $F$ was measured by a tensometric force gauge, which is part of the testing machine. Specimen extension $\Delta L$ was measured using a digital videoextensometer by X-Sight. The dependence of the force on the elongation obtained by the measurement was converted into a stress-strain diagram $\sigma$-$\varepsilon$ using the basic mechanics relations. The nominal tensile stress was determined as

$$\sigma = \frac{F}{A_0}, \tag{1}$$

where $A_0$ is the initial cross-sectional area of the specimen. The relative strain was related to the initial reference length $L_0 = 10$ mm and calculated as

$$\varepsilon = \frac{\Delta L}{L_0}. \tag{2}$$

The dependence of stress on strain in the linear region (up to the value of the proportionality limit) is expressed by Hooke's law

$$\sigma = E\varepsilon, \tag{3}$$

where $E$ is Young's modulus of elasticity in tension. In this work, the linear regression method was used to determine the Young's modulus. Stress-strain curves of ASA measured for strain rate $10^{-2}$ $s^{-1}$ and $10^{0}$ $s^{-1}$ are shown at Fig. 3.

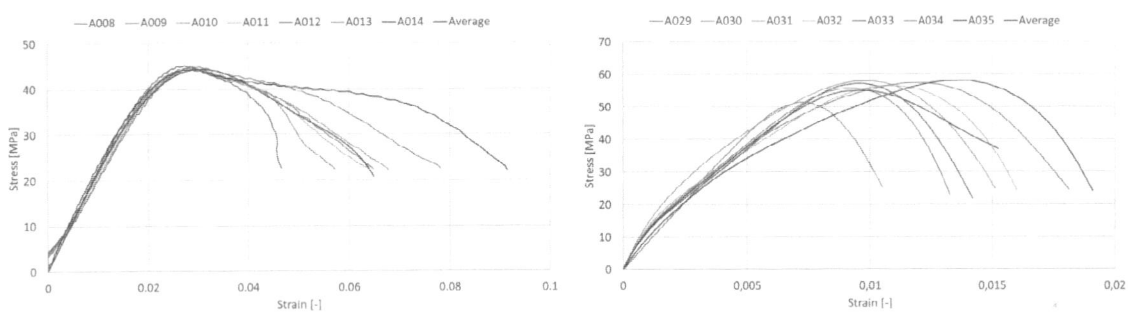


a) Tensile test results for strain rate $10^{-2}$ $s^{-1}$ b) Tensile test results for strain rate $10^{0}$ $s^{-1}$

Fig. 3: Stress-strain curves of ASA.

**Tensile tests at strain rate $10^{2}$ $s^{-1}$**

Due to the high strain rate, a drop tester was used to test the specimens. Based on the dimensions and the required strain rate, the loading rate was set at 1.2 m/s, which practically corresponds to the impact loading. The impact tensile tests were not evaluated in the same way as the tensile tests at the strain rates $10^{-2}$ $s^{-1}$ and $10^{0}$ $s^{-1}$. Unlike previous tests performed at lower speeds, in this case the elongation of the specimens was not measured. Only the force with respect to time was measured. At Fig. 4a results from impact tensile tests of ASA material are shown. The excitations that can be observed in these waveforms are due to the propagation of the shock wave from the ram impact on the lower jaw of the machine. The curves thus measured had to be adjusted after conversion to $\sigma$-$\varepsilon$ dependence using the results from the DMA and flexural tests. Matlab software, specifically the POLYFITZERO function [6], was used for the smoothing and approximation. The degree of the polynomial was chosen depending on the resulting approximation in the range from 6 to 10. An example of approximation is shown in Fig. 4b.

Since the strain rate was known ($10^{2}$ $s^{-1}$), it was possible to calculate the relative deformation $\varepsilon$ of the specimen on the basis of the time data (as function of time)

$$\varepsilon = \dot{\varepsilon} t, \tag{4}$$

where $\dot{\varepsilon}$ is strain rate and $t$ is the time until the specimen breaks. Stress-strain curves of ASA for strain rate $10^{2}$ $s^{-1}$ are shown at Fig. 4c. It should be noted that the relative deformation values obtained in this way do not fully correspond to reality and need to be corrected. The curve correction process was performed based on the results obtained from the DMA tests and a description of this procedure is not included in this paper.

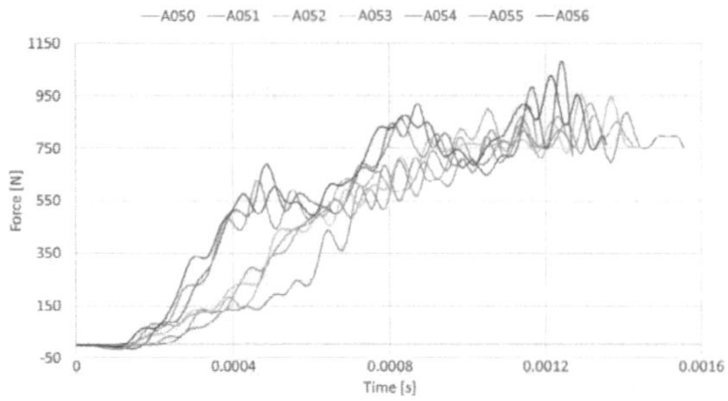


a) Tensile test results for strain rate $10^2$ $s^{-1}$

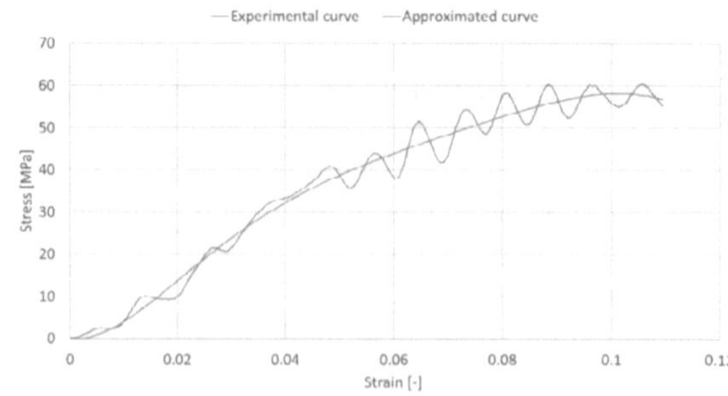


b) Approximation of measured curve

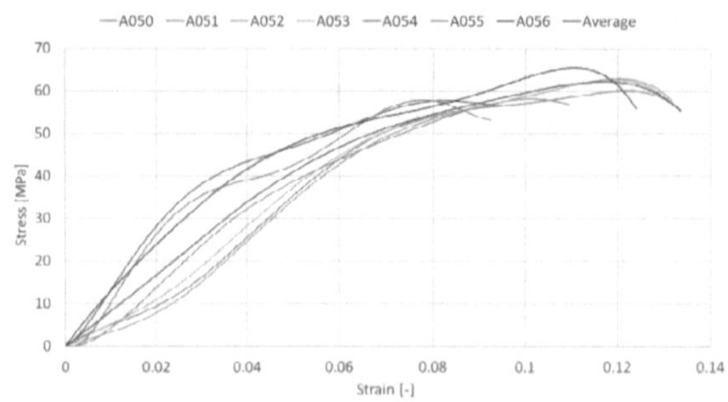


c) Approximated tensile test curves for strain rate $10^2$ $s^{-1}$

Fig. 4: Stress-strain curves of ASA and curve approximation.

From the measured curves it can be observed that the stiffness of the material increases with increasing strain rate (with higher loading speed). However, the dispersion in the elastic region also increases. At Fig. 3ait can be seen that at the strain rate $10^{-2}$ $s^{-1}$ the agreement of the curves in the elastic region is high. The strain measurement in this quasi-static tensile test is relatively highly accurate, while at higher strain rates $10^0$ $s^{-1}$ and $10^2$ $s^{-1}$ is problematic and generally less accurate. The scatter in the values of the relative deformation at failure of the specimens is also significant at all strain rates. The average value of the tensile modulus of the ASA material at the strain rate $10^{-2}$ $s^{-1}$ was found to be 2037 MPa.

## 3 ASA Material Model and Mesh Sensitivity Analysis

The evaluated area of the specimens was 10 mm long and 5 mm wide. Thus, these were dimensionally small specimens that were meshed with elements of approximately 0.5 mm in size during validation. According to the internal methodology of many automotive companies, a mesh of approximately 5 mm is used for meshing various parts. Therefore, it was necessary to investigate the effect of element size on the tensile test simulation results. For the purpose of sensitivity analysis, a special specimen model was designed, which guaranteed identical conditions for all calculations and thus better comparability of results. Compared to the original tensile specimens, this model was dimensionally four times larger, but their original cross-sectional area was retained, which means that a quarter shell thickness, i.e. 0.625 mm, was considered in the calculation. To ensure failure at approximately the middle of the specimen, the model was slightly narrowed in this section.

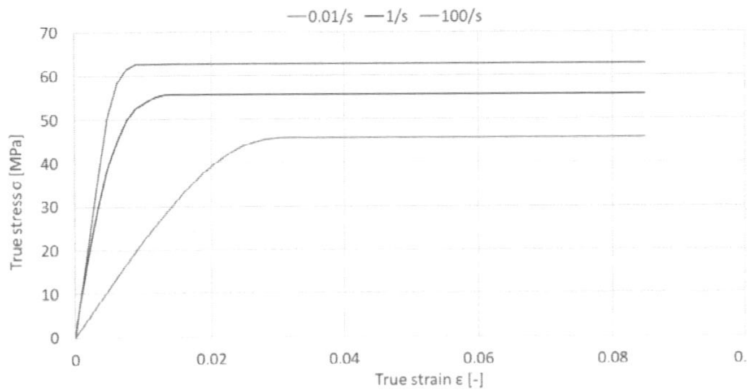


a) Stress-strain curves of ASA

b) Dimensions of the model for mesh sensitivity analysis

Fig. 5: Stress-strain curves and model for mesh sensitivity analysis.

The model was meshed with 0.5 mm, 1.5 mm, 3 mm and 5 mm elements, while the 0.5 mm mesh was considered as a reference.

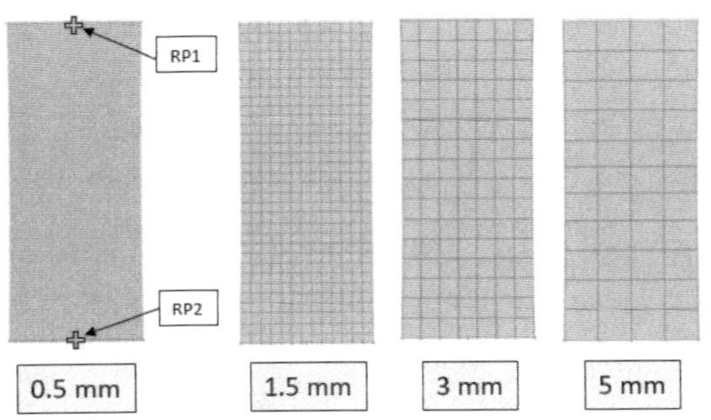


a) FEM model

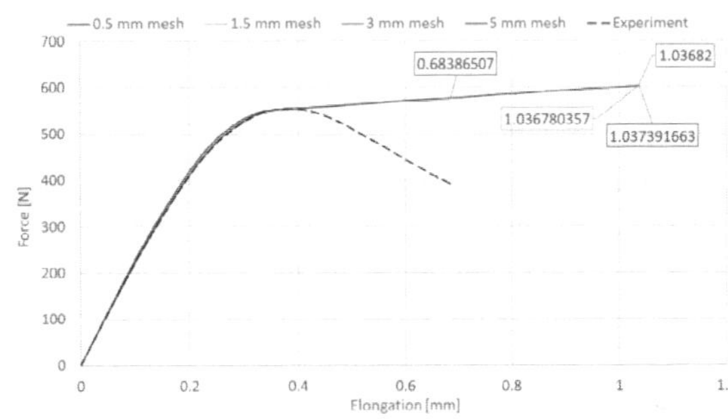


b) Results of sensitivity analysis on element size at strain rate $10^{-2}$ $s^{-1}$

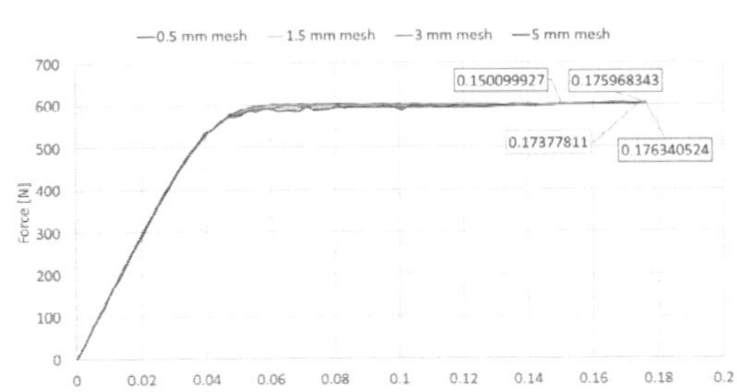


c) Results of sensitivity analysis on element size at strain rate $10^{0}$ $s^{-1}$

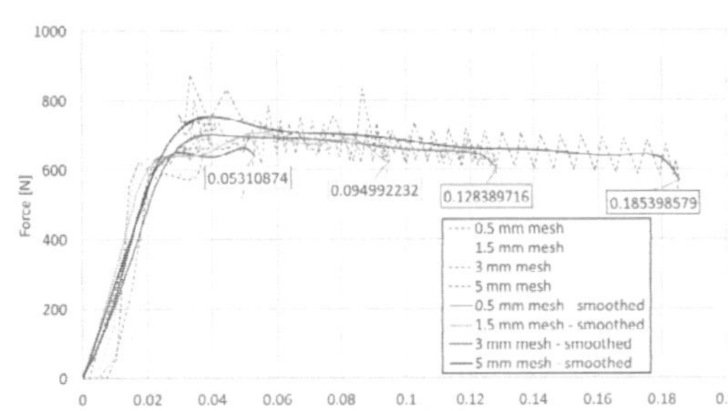


d) Results of sensitivity analysis on element size at strain rate $10^{2}$ $s^{-1}$

Fig. 6: Results of mesh sensitivity analysis.

Another fundamental difference lay in the method of deposition. In order to eliminate the possible undesirable influence of the contacts on the results of the sensitivity analysis, the sample was mounted and loaded directly in the following way. The nodes on the lower edge were connected using a multiple node to one node constraint (MTOCO) to a single reference point RP2, which was fixed, see Fig. 6a. The upper edge was connected in an identical way to the reference point RP1, in which a constant displacement velocity was defined. At Fig. 6b-d the results of the sensitivity analysis for strain rates of $10^{-2}$ $s^{-1}$, $10^{0}$ $s^{-1}$ and $10^{2}$ $s^{-1}$ are shown in the form of tensile force versus specimen elongation. It is evident from their courses that the element size does not have a significant effect on the response in the elastic region, but it does have a significant effect on the failure of the material. It was found that as the element size increases, the specimens fail later, i.e., at a higher value of the maximum elongation. For strain rates of $10^{-2}$ $s^{-1}$ and $10^{0}$ $s^{-1}$ the most significant increase in maximum elongation occurred already for the mesh with elements of 1.5 mm in size. Further enlarging of the elements did not have a significant effect on the change of its value. However, this was not the case at the highest strain rate $10^{2}$ $s^{-1}$, when the value of the elongation at the failure varied significantly with the mesh size. In view of the above findings, it was concluded that the failure curve (if defined in the simulation) obtained for the 0.5 mm element mesh must be corrected to the element size used for the automotive crash or impact simulations.

# 4 Conclusion

The results showed good agreement between simulations and experimental data, especially at lower strain rates. Simulated force-elongation curves closely followed experimental results up to the yield point. A mesh sensitivity study was performed using an enlarged specimen model. Mesh sizes of 0.5 mm (reference), 1.5 mm, 3 mm, and 5 mm were evaluated. The results showed minimal impact of mesh size on the elastic region response, but significant influence on failure prediction. Larger elements delayed failure, resulting in greater maximum elongation. Mesh size significantly affects failure predictions and must be considered when setting up simulations. These findings support the implementation of corrected material models in crash analysis scenarios and enhance the reliability of polymer simulations in technical applications.

# Acknowledgement

This article has been produced with the financial support of the projects APVV-23-0461, VEGA 1/0516/22 and European Union under the REFRESH – Research Excellence For Region Sustainability and High-tech Industries project number CZ.10.03.01/00/22_003/0000048 via the Operational Programme Just Transition.

# References

[1]. SIVIOUR, C.R., JORDAN, J.L.: High strain rate mechanics of polymers: a review, Journal of dynamic behavior of materials, Volume 2, 2016, Pages 15–32, <https://doi.org/10.1007/s40870-016-0052-8>.

[2]. FIELD, J.E. et al.: Review of experimental techniques for high rate deformation and shock studies, International Journal of Impact Engineering, Volume 30, Issue 7, 2004, Pages 725-775, ISSN 0734-743X, <https://doi.org/10.1016/j.ijimpeng.2004.03.005>.

[3]. HUANG, P.Y., GUO, Z.S., FENG, J.M.: General model of temperature-dependent modulus and yield strength of thermoplastic polymers, Chinese journal of polymer science, Volume 38, 2020, Pages 382–393, <https://doi.org/10.1007/s10118-020-2360-7>.

[4]. XIAO, X.: Dynamic tensile testing of plastic materials, Polymer Testing, Volume 27, Issue 2, 2008, Pages 164-178, ISSN 0142-9418, <https://doi.org/10.1016/j.polymertesting.2007.09.010>.

[5]. DORLÉANS, V. et al.: Time-temperature superposition in viscoelasticity and viscoplasticity for thermoplastics, Polymer Testing, Volume 101, 2021, ISSN 0142-9418, <https://doi.org/10.1016/j.polymertesting.2021.107287>.

[6]. MIKOFSKI, M.: polyfitZero. <https://www.mathworks.com/matlabcentral/fileexchange/35401-polyfitzero>.

# Detection of Surface Degradation and Damage Criteria Based on Image Processing

J.Pavlas [1,a], J. Krčil [1], S. Slavík [1], M. Růžička [1]

[1] *Czech Technical University in Prague, Faculty of Mechanical Engineering, Czech Republic*
[a] *jiri.pavlas@fs.cvut.cz*

**Abstract:** In this article, a progressive method using image processing techniques to analyse the Surface Integrity (SI) of the turbine blade surface with known load history and fatigue microcracks is proposed. The input to our proposed system consists of high-resolution images of turbine blade surfaces acquired through digital microscopy. These images are transformed into a matrix and edited to define the value of SI. The value of SI and number of flying hours is defined. These data are summarized and evaluated by a regression function. The minimal level of acceptable SI value is defined. New damage criteria are proposed based on measured data and known history. Thus, it can be predicted if the measured blade can be operated or not. This method can be generalized and used for wider application.

**Keywords:** fatigue degradation; image processing; turbine blade; turboshaft engine, nickel superalloy.

## 1 Introduction

Turboshaft engines are a hidden part of our everyday lives. They play a crucial role, especially in the power units of large helicopters, some of which are intended for rescue, firefighting and other special missions. The main parameters of these engines are their performance, operational safety and durability.

One of the most stressed components of an engine are the blades. These parts are critical, as any structural failure of a single compressor or turbine blade during operation can lead to engine failure, regardless of whether it is a stator or rotor part. Such destruction can cause secondary damage to the fuselage, Then, the helicopter can become inoperative, and this situation can lead to a catastrophic scenario. Various studies have investigated crash incidents [1], [2], [3]. It has been observed that minor changes in certain major structural and material parameters can initiate fatigue microcracks on the surface. The influence of high temperature and material degradation on the turbine blades was studied in [4][5][6][7].

It is extremely difficult to detect these potential fatigue cracks between flights or during the regular general overhaul procedure. The ability to identify and assess the level of degradation and detect the hidden fatigue cracks is essential for ensuring the safety and airworthiness of turboshaft engines. In the past, several crashes of military training and rescue helicopters caused by damage to turboshaft engine turbofan blades have occurred. The clear causes have not been found yet.

The investigation focuses on developing a systematic framework for analysing surface images to generate a quantitative fatigue assessment, thus providing an added layer of protection against potential blade destruction during flight. Image Processing Techniques used for several techniques are well known [9][10][11][12][13][14]. This principle is used to define fatigue assessment and prediction. Standard analytical techniques are well known. The real fatigue damage between the flights used for turbine blades covered by protection layer is the novelty of this research paper.

## 2 Methodology

This chapter employs image processing techniques implemented in MATLAB tool to detect and evaluate SI on turbine blades. High-resolution images of the blade surfaces were captured under controlled lighting conditions to ensure consistency and minimize reflections. The image dataset includes blades exhibiting the absence of the protective layer. Preprocessing steps involved grayscale conversion and contrast enhancement using median filtering. During operations SI value is decreasing. It means the damaged area is increasing.

Damaged area means every change of the surface after production or general overhaul. It covers mainly absence of the protective layer. This type of degradation effect initiates the first fatigue microcracks leading to the failure. The expected result of image processing is the quantification of the damaged area. It means the ratio between the damaged and undamaged area of the blade. This ratio can be expressed by following equation (dimensionless unit):

$$SI = \frac{Undamaged\ area - Damaged\ area}{Undamaged\ area} \tag{1}$$

This value, together with known number of flying hours (FH) defines the fatigue assessment which is classified as airworthy or non-airworthy blade. During operations SI value is decreasing. This means that the area of the damaged surface increases. The operational and safety concern is that some of the blades have a lower SI value than anticipated, and there is currently no method to identify the imminent risk.

The proposed procedure offers a repeatable, data-driven approach to monitoring the condition of blades. By integrating quantitative image analysis with predictive modeling, the methodology enhances early detection of surface degradation and supports more reliable maintenance planning. The procedure consists of following steps:

1) Sample Preparation: Blades are cleaned thoroughly and re-coated with a new, standardized protective layer according to the maintenance manual.
2) Lighting Configuration: Controlled and uniform lighting conditions are established to reduce the influence of shadows, glare, and reflective distortions inherent to metallic surfaces. This ensures reproducibility and accuracy of image capture.
3) Image Acquisition: Each blade is photographed under the defined lighting setup to generate high-quality, consistent images suitable for further analysis.
4) Image Preprocessing: The acquired images undergo grayscale conversion and contrast enhancement to improve visibility of surface features and degradation patterns.
5) SI Analysis: Calculating the black-to-white pixel ratio within each image, which serves as an indicator of surface degradation. A minimum of 30 blade samples should be analysed to produce statistically robust results.
6) Data Collection: SI values are plotted against the corresponding FH for each sample, generating a dataset that captures the degradation progression under real operational conditions. SI value is decreasing with increasing FH value. It means the higher FH, the higher surface degradation and smaller SI. For blades with approximately more than 1000 FH, the SI typically decreases to about 0.4.
7) Regression Analysis: A best-fit regression curve is derived from the dataset to model the typical degradation trend over time. This function serves as the foundation for establishing predictive benchmarks.
8) Definition of Safety Threshold: To ensure reliability and airworthiness, a minimum acceptable SI threshold is defined by dividing the regression curve by a chosen safety factor (e.g., 1.5). This accounts for variability in usage conditions and manufacturing tolerances.
9) Airworthiness definition: A degradation map is constructed with two distinct zones:
   - Area 1: Blades with SI values above the minimum threshold are classified as airworthy.
   - Area 2: Blades falling below the threshold are considered non-airworthy and recommended for further inspection or replacement.
10) SI Classification of the new blade: The SI value obtained from the processed image is compared to the predefined threshold curve. Based on this comparison, the blade is classified into Area 1 (airworthy) or Area 2 (non-airworthy), enabling fast, non-destructive decision-making during maintenance procedures.

# 3 Example

This chapter describes the experimental program focused on the selected sample. For this experiment, the second stage turbocompressor rotor blade of the helicopter turboshaft engine TV3-117 was chosen as a sample (Fig. 1). The turbocompressor consists of two stages blade assembly, where hot gases from the gas chamber are released.

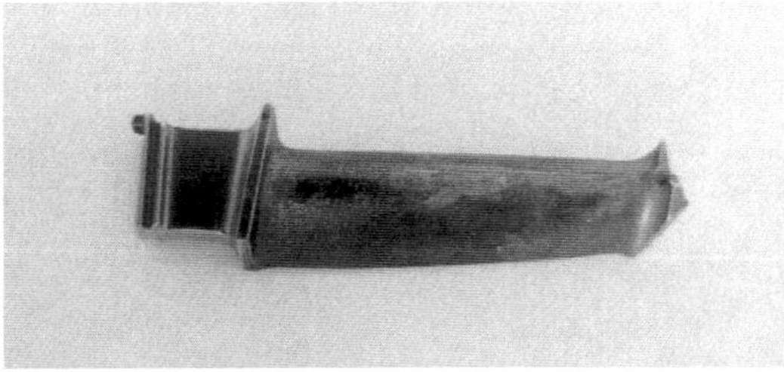

Fig. 1. The demonstrator - turbocompressor blade.

This stage consists of 101 rotor blades. Structural design, operational and maintenance data of TV3-117 are summarized in [15]. The blade structure consisted of two different materials. Nickel alloy ZS6K is substrate material of the blade covered by a protective aluminium layer. Such designed blades can withstand aerodynamic pressure, the centrifugal force, high temperature and exhaust gases. The chemical composition is defined in Table 1.

Table 1: Chemical composition in [%] of nickel alloy ZS6K [16]

| C | Cr | Al | Ti | Mo | Co | W | Fe | Mn<br>Si | P<br>S | Ce | B | Pb | Bi |
|---|---|---|---|---|---|---|---|---|---|---|---|---|---|
| 0.13-0.2 | 9.5-12 | 5-6 | 2.5-3.2 | 3.5-4.8 | 4-5.5 | 4.5-5.5 | 2 | 0.4 | max<br>0.015 | max<br>0.015 | max<br>0.02 | max<br>0.005 | max<br>0.0015 |

The basic operational engine data are collected in Table 2. The nominal revolution value (100%) is 19500 RMP. There are five basic operational regimes with defined maximum allowed RPM and temperature values.

Table 2: Operational regimes, revolution and temperatures before the turbocompressor

| Regime | Maximum allowed revolutions [% of 19500 RPM] | Maximum allowed engine sensor temperature [°C] |
|---|---|---|
| Emergency | 101.15 | 990 |
| Take off | 101.15 | 990 |
| Nominal | 99 | 955 |
| Cruise 1 | 97.5 | 910 |
| Cruise 2 | 95.5 | 870 |

The experiment involves the collection and classification of data from 30 TV3-117 turboshaft engines, each contributing a randomly selected turbocompressor blade that had been subjected to real operating conditions in both civil and military training environments. The monitored data focuses on the trailing edge region located at the mid-span of the blade, identified through maintenance records as the most frequently damaged area.

All engines were disassembled in accordance with the procedures defined in the general overhaul manual, and blade examinations were carried out as part of standard maintenance protocol. All engines were operated in the interval of 0-1500 flying hours.

### 3.1 Input data

Fig. 2 shows the surface degradation after the operation (1325 FH) and before the mandatory general overhaul. The blade surface is full of specific non-homogeneous craters. Maximal observed diameter of these craters is 20μm with maximal depth 10μm. These craters connect to each other and form larger areas. Digital microscopy analyses were performed with using digital optical microscope Olympus DSX1000. Then the images surface quality was analysed by image processing technique using MATLAB tool Fig. 3.

The white areas in Fig. 3 represent the residual parts of the protective layer. The black areas show substrate material without any protection. This means that the substrate material is exposed to gas exhaust directly. SI value of this blade is defined as 0.38 which means that this blade is at the end of TBO.

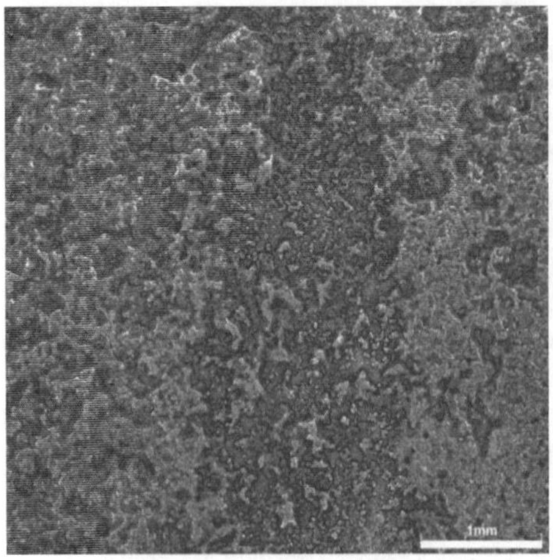


Fig. 2. Unedited image.

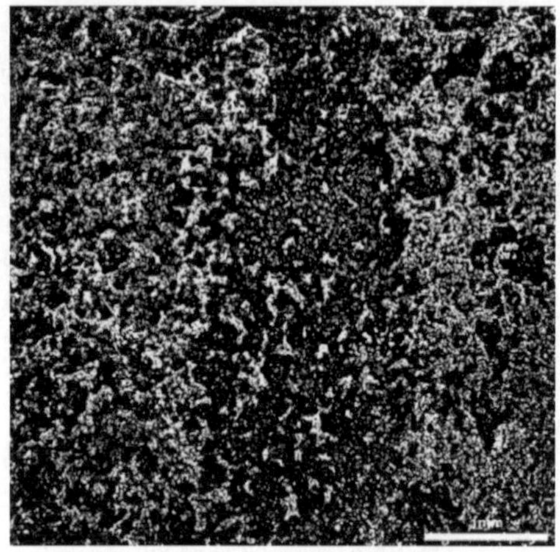

Fig. 3. Edited image.

Example of the cross section of the blade with degradation shows Fig. 5. The protective layer is damaged and high-oxidation degradation can grow with potential of the first initial fatigue crack.

Fig. 4. Cross section of the damaged blade in detail.

### 3.2 Data regression and classification

SI values were obtained from a dataset comprising 30 TV3-117 turbocompressor blades before GO and plotted with respect to operational FH. A linear regression analysis was performed to characterize the degradation trend and establish a predictive model of SI decline over time. To define a conservative damage criterion, the linear regression curve was divided by a safety factor of 1.5, resulting in a minimum acceptable SI threshold. Blades were subsequently classified based on their SI values relative to this threshold Fig. 5. This chart shows the relation between increasing FH and decreasing SI. Blades exceeding the threshold were deemed airworthy, while those below were categorized as non-airworthy.

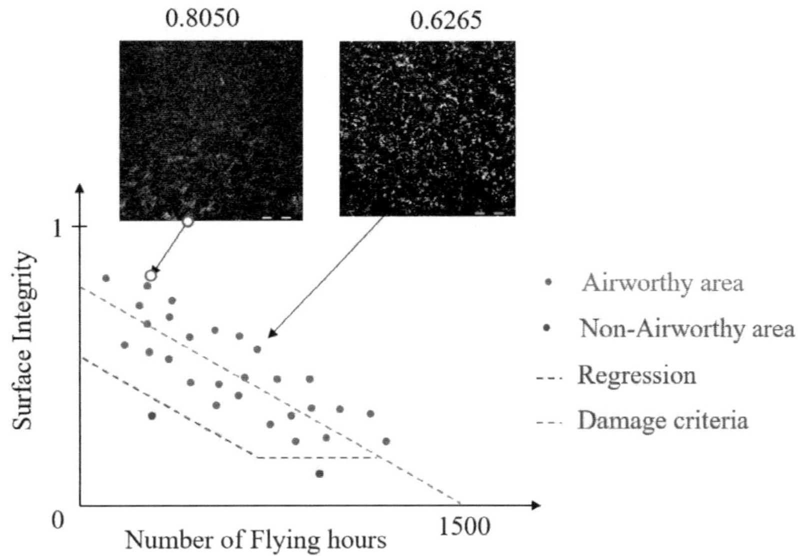

Fig. 5. Data regression and classification.

## 4 Results

The observed data show the relation between SI and FH. New damage criteria are proposed based on these values. This method can be used for the predictive maintenance system. This method can also be transformed into a standard service procedure between flights if borescope techniques are available and adjustable. SI values can be influenced by the load spectrum and environmental conditions. If a helicopter is used for a training, engines are exposed to a higher load.

Surface degradation and fatigue microcracks were identified in investigated blades, primarily concentrated in the leading-edge region. This area, characterized by minimal thickness, is particularly vulnerable to thermomechanical stresses induced by rapid cooling during immediate engine shutdowns. Such shutdowns, common in training exercises, emergency landings, or military operations, result in non-uniform thermal gradients that provoke localized stress peaks and deformation. If the protective layer thickness is below 0.02 mm, the coating loses its capacity to shield the substrate, leading to thermal degradation and burnout. Conversely, if the coating exceeds 0.05 mm, partial delamination may occur, leaving the substrate material unprotected.

Additionally, surface roughness of the protective layer contributes significantly to fatigue crack initiation. Microcracks frequently nucleate at geometric surface irregularities (as shown in Fig. 2), which act as local stress concentrators. Reducing these irregularities presents a technological challenge, as achieving smoother coatings without compromising other material properties, remains difficult within existing manufacturing constraints.

Another influence is the presence of dust, sand and other debris. These small particles can seriously damage the protective layer of the blade. From the design organisation, known operations and findings from the maintenance organisation, it can be defined how SI can decrease in detail.

Notably, even new blades exhibit an initial SI value of approximately smaller than 1 (e.g., 0.8) due to manufacturing marks and reflective areas while the ideal baseline is defined as 1.0.

## 5 Conclusion

The turbine blades of the turboshaft engine TV3-117 were subjected to the proposed method based on image processing, linear regression and data classification. The relation between surface degradation, operation and hidden fatigue microcracks was determined. The results of observed data were used for the proposed new damage criteria.

The proposed method demonstrated significant potential in accurately predicting level of degradation by effectively capturing complex nonlinear relationships inherent in engine operational, maintenance and production data. This methodology not only increases the reliability of fatigue assessment, but also offers a computationally efficient alternative to traditional engineering techniques. The image processing approach provides a non-destructive, scalable methodology for condition monitoring and offers potential for integration into predictive maintenance systems.

The industrial aspects of this research will help detect potentially dangerous engines. The identification of these engines helps to increase the level of aviation safety and decrease the number of future aircraft accidents.

Moreover, this method can help predict the necessity of spare parts and simplify the logistic procedures.

Future steps will be focused on an in-depth description of surface degradation effects, and analytical and numerical methods for predicting their growth.

## Acknowledgement

This publication was supported by the Czech Technical University in Prague, Faculty of Mechanical Engineering, Czech Republic.

## References

[1] A.M. Kolagar, et al., Failure analysis of gas turbine first stage blade made of nickel-based superalloy, Case Stud. Eng. Fail. Anal. 8 (2017) 71–78, https://doi.org/10.1016/j.csefa.2017.04.002.

[2] R.K. Mishra, et al., Failure analysis of an un-cooled turbine blade in an aero gas turbine engine, Eng. Fail. Anal. 79 (2017) 836–844, https://doi.org/10.1016/j.engfailanal.2017.05.042.

[3] Z. Khan, et al., Investigation of intergranular corrosion in 2nd stage gas turbine blades of an aircraft engine, Eng. Fail. Anal. 68 (2016), https://doi.org/10.1016/j.engfailanal.2016.05.033.

[4] J. Błachnio, J. Spychała, D. Zasada, Analysis of structural changes in a gas turbine blade as a result of high temperature and stress, Eng. Fail. Anal. 127 (2021) 105554, https://doi.org/10.1016/j.engfailanal.2021.105554.

[5] Bartošák, M., et al., High-temperature low-cycle fatigue and fatigue–creep behaviour of Inconel 718 superalloy: Damage and deformation mechanisms, International Journal of Fatigue, ISSN: 0142-1123, Vol: 186, Page: 108369, 2024. https://doi.org/10.1016/j.ijfatigue.2024.108369

[6] T.S. Chowdhury, et al., A critical review on gas turbine cooling performance and failure analysis of turbine blades, Int. J. Thermofluids 18 (2023) 100329, https://doi.org/10.1016/j.ijft.2023.100329.

[7] J. Kanesund, H. Brodin, S. Johansson, Hot corrosion influence on deformation and damage mechanisms in turbine blades made of IN-792 during service, Eng. Fail. Anal. 110 (2020) 104388, https://doi.org/10.1016/j.engfailanal.2020.104388.

[8] B. Fu, et al., Crystal plasticity modeling of fretting fatigue crack initiation behavior in Ti6Al4V, Eng. Fail. Anal. 165 (2024) 108663, https://doi.org/10.1016/j.engfailanal.2024.108663.

[9] G.K. Ijemaru, et al., Image processing system using MATLAB-based analytics, Bull. Electr. Eng. Inform. 10 (2021) 2566–2577, https://doi.org/10.11591/eei.v10i5.3160

[10] Krzysztof Pałczyński, An image processing approach for fatigue crack identification in cellulose acetate replicas, Engineering Failure Analysis, Volume 164, October 2024, https://doi.org/10.1016/j.engfailanal.2024.108663

[11] K. E. Hazzan, M. Pacella, Crack Identification in Tungsten Carbide Using Image Processing Techniques, Procedia Structural Integrity, Volume 37, 2022, Pages 274-281, https://doi.org/10.1016/j.prostr.2022.01.085

[12] J. Li et al., A Machine Vision Method for Detecting Pineapple Fruit Mechanical Damage, Agriculture 2025, 15(10), 1063; https://doi.org/10.3390/agriculture15101063

[13] J. Chen et al, Research on Tire Surface Damage Detection Method Based on Image Processing, Sensors 2024, 24(9), 2778; https://doi.org/10.3390/s24092778

[14] B. Fu, et al., Crystal plasticity modeling of fretting fatigue crack initiation behavior in Ti6Al4V, Eng. Fail. Anal. 165 (2024) 108663, https://doi.org/10.1016/j.engfailanal.2024.108663.

[15] M. Furda, Vrtulnik Mi-17, Velitelstvi vycviku – Vojenska akademie Vyskov, 2016.

[16] J. Belan, Quantitative metallography of heat treated ZS6K superalloy, Mater. Eng. - Mater. Inžinierstvo 18 (2011) 121–128.

# Performance Optimization of CFRP Retaining Sleeve for Rotors via Controlled Filament Winding Pre-Stress: Experimental Approach

O. Pavlík [1,a], T. Návrat [1], G. Cabaj[1], M. Vajdák[2]

[1] Institute of Solid Mechanics, Mechatronics and Biomechanics, Faculty of Mechanical Engineering, Brno University of Technology, Czech Republic.
[2] Inpraise Systems s.r.o.
[a] ondrej.pavlik1@vutbr.cz

**Abstract:** This paper investigates the effect of controlled pre-stress applied during the filament winding of carbon fiber reinforced polymer (CFRP) retaining sleeves, intended for high-speed permanent magnet synchronous machine (PMSM) rotors. Using pre-preg unidirectional carbon fibers, a set of test specimens was fabricated under varying tension loads to simulate realistic rotor conditions. Tensile tests, supported by digital image correlation (DIC), were performed to evaluate strain distribution and mechanical integrity. The results demonstrate a clear correlation between the applied pre-stress and enhanced strength, stiffness, and homogeneity of the CFRP structure. The findings contribute to better material utilization and rotor durability in demanding application, while mitigating the risk of critical failure scenarios during rotor operation.

**Keywords:** PMSM, CFRP, Filament winding, Pre-stress, tensile testing, DIC

## 1 Introduction

The continuous advancement of electric motor technology, especially for high-speed applications, requires materials that provide high mechanical strength, lightweight construction, and reliable performance under extreme loading. Permanent magnet synchronous machines (PMSMs) benefit from carbon fiber reinforced polymer (CFRP) retaining sleeves [3], which ensure magnet retention at high circumferential speeds. In addition to its mechanical advantages, the composite material exhibits a lower density, which reduces centrifugal forces at high rotational speeds [1]. Furthermore, its non-magnetic nature fulfills the critical requirement of electromagnetic compatibility in the gap between the rotor's permanent magnets and the stator winding [1].

Fig. 1: Pre-preg composite tow, before winding process and before thermal curing process.

High-speed PMSM rotors require CFRP retaining sleeves to ensure magnet retention, provide sufficient stiffness to avoid critical speeds, and minimize the rotor–stator gap for electromagnetic efficiency. All of these functions are influenced by the applied winding pre-stress [2].

One of the critical parameters influencing the mechanical and functional quality of the filament-wound retaining sleeve is the level of pre-stress applied during the manufacturing process. While it may seem intuitive to apply the maximum possible tension to the fibers during winding in order to enhance compressive preload and structural integrity, this approach can be counterproductive [4]. Excessive pre-stress may induce micro-damage within the fiber tow—comprising thousands of individual filaments—which may not be visually detectable but can significantly compromise the overall mechanical performance of the composite [5]. Thus, a balanced application of pre-stress is essential to ensure optimal fiber alignment and interfacial bonding without introducing defects. This represents a fundamental limitation of current theoretical and computational approaches. The accumulation of microstructural damage throughout the manufacturing chain—including fiber tow handling, pre-stress application, filament winding-both machine and operator, and thermal cycling during curing—introduces complex degradation mechanisms that are not yet reliably captured by existing analytical or numerical models [6]. Although the applied pre-stress can be accurately measured and readily implemented in finite element simulations for rotor design and margin of safety (MoS) estimation, these models typically assume ideal, undamaged material behavior. Neglecting the effects of process-induced defects and stress concentrations can therefore lead to a significant overestimation of mechanical performance and an artificially inflated safety margin, potentially compromising the reliability of the final design.

# 2 Objectives and Experimental Strategy

## 2.1 Research Objectives

The primary objective of this study is to experimentally assess the influence of filament winding pre-stress on the mechanical properties of CFRP retaining sleeves, intended for use in high-speed permanent magnet synchronous machine (PMSM) rotors. The motivation stems from the need to identify an optimal pre-stress level that maximizes structural integrity without compromising the internal fiber architecture.

Specifically, this study seeks to establish a quantifiable relationship between applied pre-stress and key material characteristics such as tensile strength, strain uniformity, and failure behavior, thereby informing rotor design practices and enhancing material utilization. Furthermore, the process must be fully re

plicable to allow for direct comparison between different material candidates, and above all, aims for simplicity and cost optimization due to the high cost of manufacturing rotors and optimizing via destructive rotational tests.

Fig. 2: Mock-up "magnets" for filament winding process.

## 2.2 Methodological Approach

Pre-preg unidirectional carbon fiber tows were selected (Fig. 1) as the base material in order to minimize variability typically associated with resin-rich zones and fiber wetting, which are common in wet-winding methods [2]. The mechanical behaviour was assessed via quasi-static tensile testing, supported by full-field surface strain analysis using digital image correlation (DIC). This combination provided both global strength data and high-resolution local deformation mapping. The methodology allowed for direct correlation between pre-stress levels applied during manufacturing and resulting strain fields under load, capturing subtle effects of internal defects or misalignment.

## 2.3 Specimen Design and Manufacturing

Due to the extreme forces involved in full-scale rotor testing, a scaled tensile specimen was developed (Fig. 3) that preserved the essential mechanical and geometric characteristics of the CFRP retaining sleeve. Each specimen was wound on a rigid mandrel incorporating mock inserts simulating permanent magnets (Fig. 2 & Fig. 3), in order to replicate the boundary conditions, present during actual rotor winding.

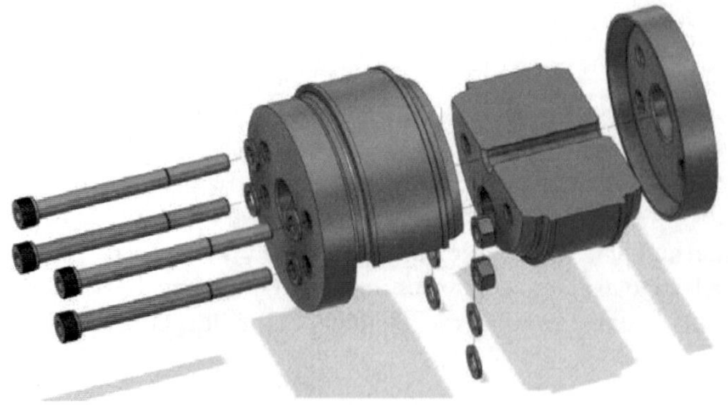

a) Set-up for filament winding, 2x "magnets" mock-up and supporting flanges bolted together.

b) Compiled set-up for filament winding process.

Fig. 3: Set-up for filament winding process including magnet mock-ups and supporting flanges with connecting bolts.

Pre-stress was applied to the filament tows in discrete steps, ranging from 5 kg to 30 kg in 5 kg increments. After winding, the specimens were cured thermally according to the manufacturer's guidelines for the pre-preg system. Post-curing, all samples underwent precision machining to achieve a uniform thickness of 0.8 mm (Fig. 4 a), ensuring consistent stress distribution during tensile testing and ensuring consistent surface conditions (roughness, tolerances, machining), comparable to those found in operational high-speed rotors.

a) Composite detail after flanges removal and machining operation.

b) Manufactured specimens after filament winding and curing process, before flanges removal.

Fig. 4: Manufacturing of test specimens after winding process, curing process and flanges disassembly.

A total of thirteen specimens (Fig. 4 b) were manufactured and tested in one batch, experimental campaign was repeated four times, with up to four specimens from each campaign being disqualified due to visible failures during winding/curing process, or errors during machining and finishing, reducing the overall number of samples but conforming to quality criteria of high-speed capable rotors mock-up. Particular attention was paid to visual inspection, identification marking, and handling throughout the process to eliminate confounding variables and ensure reproducibility of results.

# 3 Experimental Setup and Testing Procedure

The mechanical properties of the specimens were evaluated using a servo-hydraulic universal testing machine with a maximum load capacity of 250 kN. Custom flat wedge grips were employed (Fig. 5 a) to ensure proper load transfer and minimize slippage or stress concentrations at the specimen ends. Each specimen was loaded in uniaxial tension until complete failure. Testing was conducted under displacement control at a constant loading rate suitable for quasi-static conditions.

a) Custom manufactured grips able to withstand estimated forces during the destruction of the CFRP.

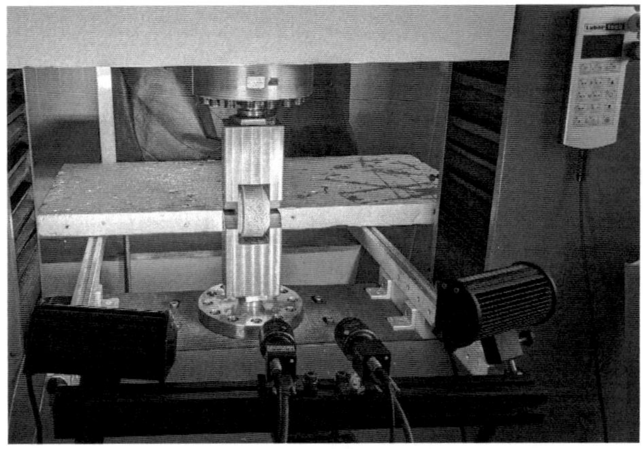

b) Experimental set-up with tensile test machine and DIC camera system.

Fig. 5: Set-up for stress-strain testing and DIC measurement.

In addition to conventional force–displacement measurement, high-resolution full-field surface strain data were acquired using a DIC system (Fig. 5 b). The specimens were prepared with a stochastic speckle pattern applied to the outer surface prior to testing. Two synchronized high-speed cameras captured the deformation field throughout the test, allowing post-processing of strain distribution and localization behaviour up to the point of fracture.

The goal was to obtain maximum stress during failure and correlate with applied pre-stress to obtain the % of material properties reduction due to fiber damage, DIC analysis provided additional insight into strain homogeneity, strain localization, and any early signs of damage initiation.

Data were processed using commercial DIC software, enabling the generation of two-dimensional strain maps across the specimen surface. These were used to verify consistency between the applied pre-stress levels and resulting mechanical response. All measurements were conducted under controlled laboratory conditions, and results were statistically evaluated to establish trends and quantify variability between test groups.

# 4 Results

This chapter presents the experimental findings obtained from the tensile testing of filament-wound CFRP specimens produced under varying pre-stress levels. The results include both conventional force–displacement data and detailed strain field measurements captured through Digital Image Correlation (DIC).

It is evident, that the CFRP exhibits no noticeable plastic deformation and its fracture behavior is characterized by a brittle failure mode (Fig. 6).

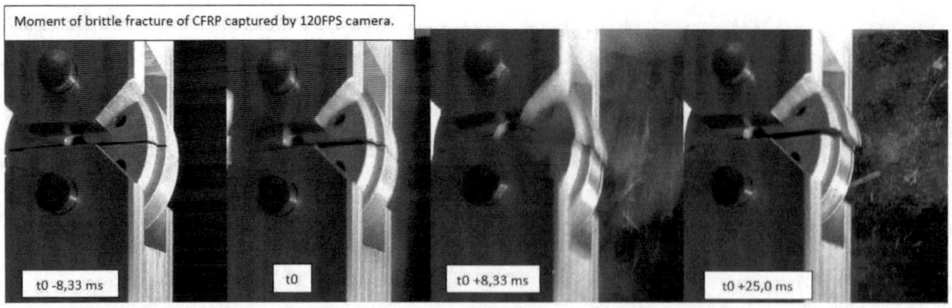


Fig. 6: Moment of brittle fracture of the CFRP captured by 120FPS camera.

Force-Displacement curves are obtained (Fig. 7), while the displacement encompass also the deformation of the vices, the Force at break and the overall behavior are the driving factors. These outputs are analyzed to evaluate tensile strength, stiffness, and strain uniformity (Fig. 8), with particular attention given to the effects of pre-stress on material behavior under load. The following figures and tables illustrate the mechanical response of the specimens and highlight key trends and anomalies observed during testing.

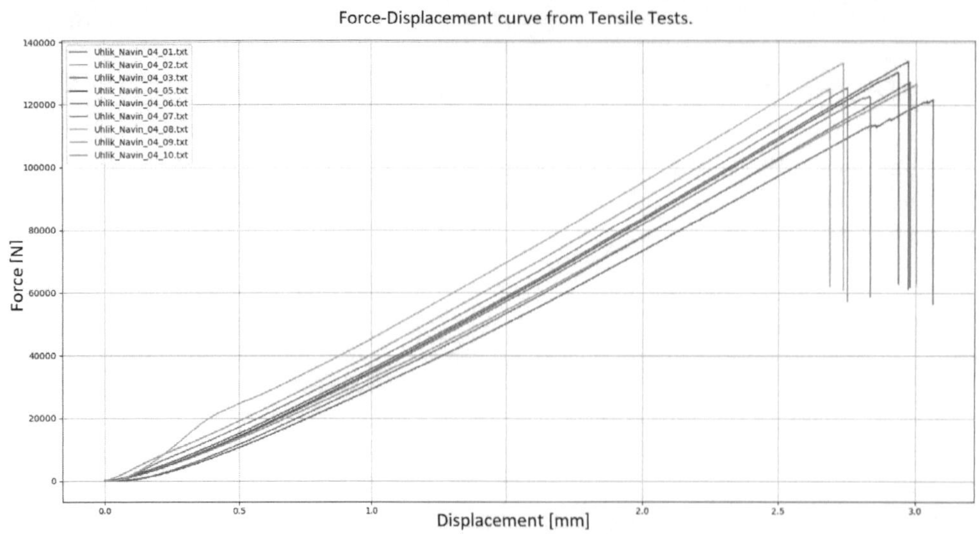


Fig. 7: Force-Displacement curve.

Shift in the material's strength limit was evaluated, absolute limit was established at 2820 ± 87.4 MPa with 25kg pre-stress on the fiber (Tab. 1), which was the best candidate for the selected pre-preg (Toray fiber). Slippage and other negative phenomena observed in the initial tests were successfully eliminated with 25 kg pre-stress. Additionally, the Digital Image Correlation (DIC) results agree with the expected material behavior.

Tab. 1: Experimental results of a test campaign #3 and #4

| **Pre-Stress [kg]** | **5** | **10** | **15** | **20** | **25** | **30** |
|---|---|---|---|---|---|---|
| quantity | 2 | 2 | 2 | 2 | 10 | 1 |
| Fmax[1] [kN] | 121.6; 126.5 | 127.1; 130.3 | 122.6; 133.9 | 125; 125.5 | <124; 144.8> | 133 |
| $\sigma_{UTS}$[2] [MPa] | 2533; 2635 | 2648;2715 | 2554; 2790 | 2604; 2614 | <2583; 3017> | 2771 |

[1]*Experimental*
[2]*Calculated*

Strain distribution according to DIC was uniform across the specimen throughout the tensile test. Strain values before failure ranged from 0.8% to 1.5%. According to the datasheet, the maximum strain for the carbon fibers is 2%, suggesting that the strain might be higher on the inner side and decreases closer to the surface due to the specimen's rounded profile.

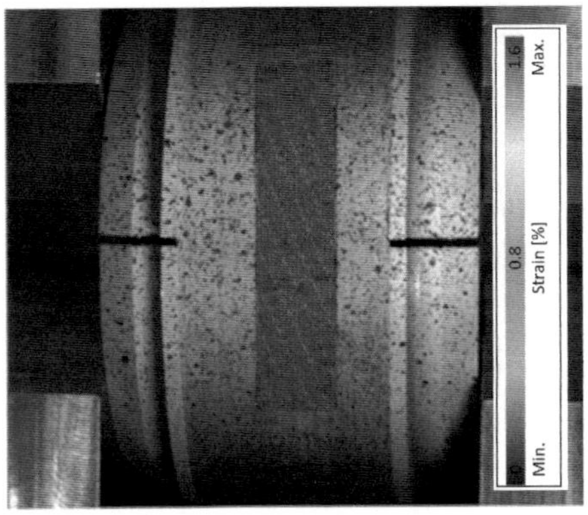


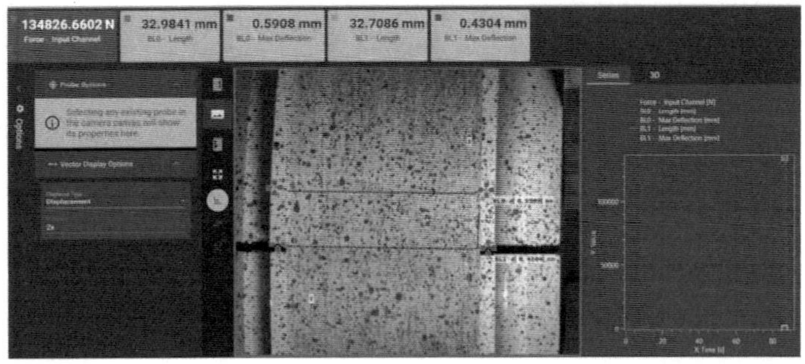


a) Strain distribution on the surface of the specimen.

b) „Bend-line", displacement evaluation on the surface layer.

Fig. 8: DIC (Digital Image Correlation) Evaluation: Strain distribution, Bend line and detail: Looking for artefacts or any visible errors within the distribution and possible correlation with pre-stress.

Displacement throughout the entire tensile testing process was relatively uniform, with local extremes occurring near the specimen edges, where boundary constraints tend to induce interaction effects. This behavior is clearly illustrated in Fig. 9. In some cases, these edge effects led to localized delamination of the fiber tow (Fig. 9 b); however, due to the limited extent and confinement of the affected regions, no measurable impact on the overall mechanical performance was observed.

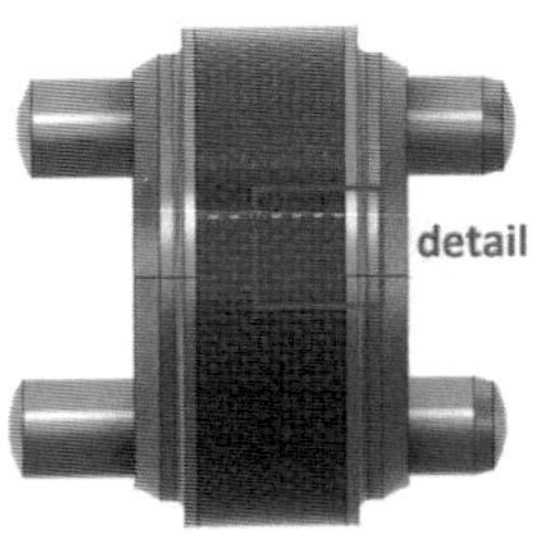


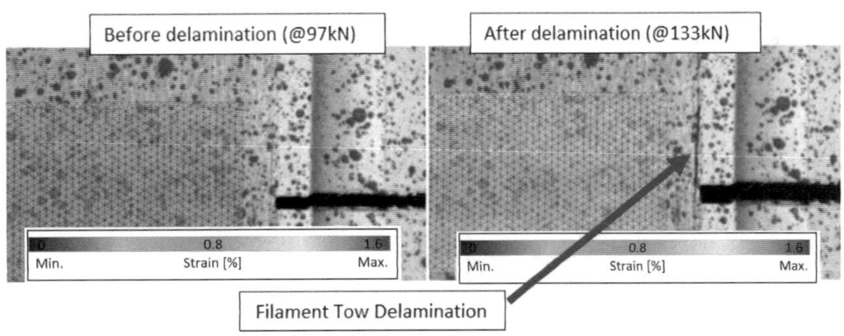


a) Localized area search

b) Single pre-preg tow delamination visible.

Fig. 9: Only visible case of delamination occurring, at this amount not considered a major fault.

## 5 Discussion

The analysis of experimental results revealed two contrasting failure mechanisms associated with extreme values of applied pre-stress during filament winding. These findings are critical for guiding manufacturing parameters and ensuring reliable rotor behavior under real-world operating conditions.

For specimens produced with very low pre-stress levels (≤ 5 kg), the stress–strain curves display nonlinear behavior and abrupt discontinuities (Fig. 10). This irregular response appears beyond ~60% of the material's ultimate tensile strength and is indicative of delayed load-sharing within the fiber tow structure. The likely cause is fiber slippage due to insufficient compaction during winding. As the load increases, the loosely packed fibers gradually engage, which may cause stepwise stiffness increases or sudden redistributions of internal stresses. While this behavior may not lead to immediate failure under static conditions, its unpredictability under dynamic loading poses a significant risk to rotor integrity—particularly in high-speed applications where even minor imbalances can lead to severe consequences.

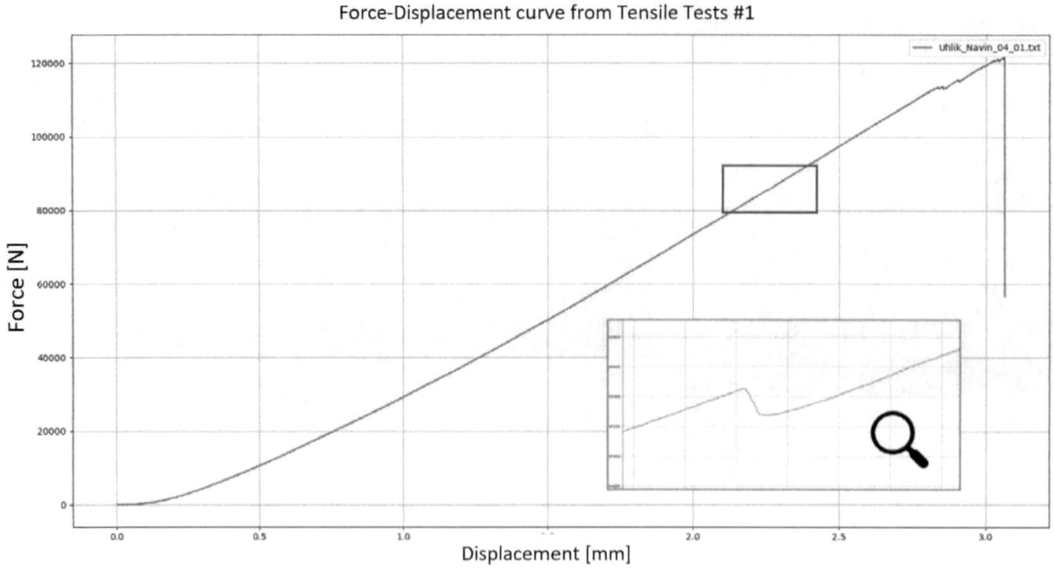


Fig. 10: Behavior of filament wind with pre-stress 5 kg (or less).

In contrast, specimens produced with the highest pre-stress levels (≥ 30 kg) exhibited high ultimate strength but also non-uniform stiffness development, as illustrated in Fig. 11. Rather than a purely linear response, the force–displacement curve shows a distinct shift in slope in the early loading phase, indicating changes in stiffness as the load progresses.

This nonlinear behavior may be a result of partial internal damage within the fiber tow—such as micro-buckling, resin micro-cracking, or uneven stress transfer between filaments—which becomes active as the material is increasingly loaded. While the specimen ultimately sustains high force, these irregularities suggest the internal stress distribution is suboptimal and may evolve during loading. In dynamic applications, such evolving stiffness could compromise rotor stability or induce unanticipated vibrations.

The data also show that as pre-stress levels increase, the likelihood of producing defective specimens rises, likely due to the narrower margin for manufacturing error under higher tension.

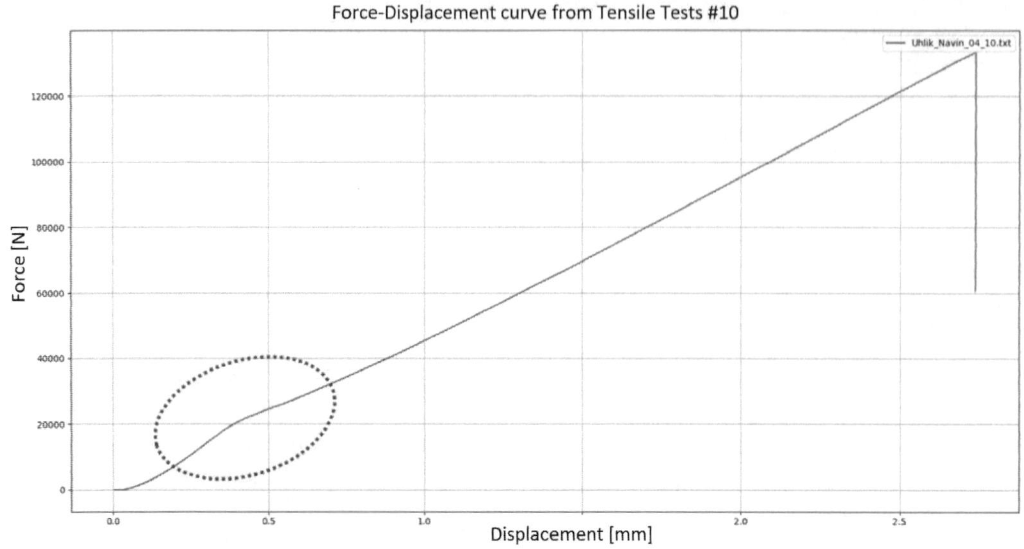


Fig. 11: Behavior of filament wind with pre-stress 30kg or more.

Moreover, the data suggest an increasing rate of manufacturing defects at higher winding tensions. While all specimens were visually inspected and machined to consistent tolerances, the rejection rate rose with higher pre-stress levels, likely due to the increased sensitivity of the process to minor handling errors or local material imperfections.

Importantly, such inconsistencies can be filtered out in static tensile tests, but might not be detectable during actual rotor winding, posing a serious risk to quality control.

## 6 Conclusion

This study demonstrates that the application and optimization of pre-stress during the filament winding process with the goal of enhancing the mechanical performance of carbon-fiber composite retaining sleeves for high-speed PMSM rotors.

Laboratory tests indicate that such pre-stress improves tensile strength, uniform stress distribution, and overall fiber alignment, which are critical for rotor durability and reliability, however attention must be given to the negative phenomena observed with either too low or too high pre-stress.

Although stress-limit of the fiber is relatively simple to experimentally optimize, it is much harder task to avoid all negative effects correlating with too high or too low pre-stress, for which there is currently not an effective analytical or numerical approach to predict and mitigate.

Further research, particularly using full-scale rotor test rigs is needed to adequately describe and understand all of these complex mechanics. Especially within dynamic environment, as the next frontier.


## Acknowledgement

This work/publication was supported by the project "Partnership for Security in Sustainable Transport and Energy (PSSITE)", funded as project No. CZ.02.01.01/00/23_020/0008549 by Programme Johannes Amos Commenius, call Intersectoral Cooperation.


## References


[1] J. Barbero, *Introduction to Composite Materials Design*, 2nd ed., Boca Raton: Taylor & Francis, 2011, ISBN 9781420079159.

[2] A. Błachut, T. Wollmann, M. Panek, M. Vater, J. Kaleta, J. Detyna, S. Hoschützky, M. Gude, Influence of fiber tension during filament winding on the mechanical properties of composite pressure vessels, Composite Structures, 304 (2023) 116337, https://doi.org/10.1016/j.compstruct.2022.116337.

[3] Y. Duan, D. M. Ionel. A Review of Recent Developments in Electrical Machine Design Optimization Methods With a Permanent-Magnet Synchronous Motor Benchmark Study. IEEE Transactions on Industry Applications, vol. 49, no. 3, pp. 1268–1275, June 2013.

[4] O. Abdullah, A. Hassan, Effect of prestress level on the strength of CFRP composite laminate, Journal of Mechanical Science and Technology 30 (11) (2016) 5115–5123, https://doi.org/10.1007/s12206-016-1029-1.

[5] E. Toptaş, N. Akkuş, Damage detection of carbon fibers in filament winding machines using an electrical resistance method, The International Journal of Advanced Manufacturing Technology 93 (2017) 671–679, https://doi.org/10.1007/s00170-017-0457-0.

[6] X. Chen, Y. Li, D. Huan, H. Liu, L. Li, Y. Li, Influence of filament winding tension on the deformation of composite flywheel rotors with H-shaped hubs, Materials 15 (6) (2022) 2136, https://doi.org/10.3390/ma15062136.

# Experimental Investigation of PA6 GF under Tensile and Relaxation Loading

J. Prajka [1,a], R. Halama [1], M. Šofer [1], T. Vlček [1]

[1] 1 *Department of Applied Mechanics, Faculty of Mechanical Engineering, VSB - Technical University of Ostrava, Ostrava, Czech Republic*

[a] *jaroslav.prajka.st@vsb.cz*

**Abstract:** Polyamide 6 reinforced with glass fibers (PA6 GF, Grivory) is widely used in structural applications where time-dependent deformation can influence long-term performance. This study investigates the mechanical response of PA6 GF using tensile and creep tests. Tensile tests were conducted at multiple strain rates to evaluate the material rate sensitivity, while relaxation tests were performed to assess the gradual reduction of stress over time. A stepwise deformation approach was applied to observe transitions in relaxation behavior. Experimental data provided insight into the viscoplastic response of PA6 GF and the strain levels at which relaxation effects become significant. In addition to experimental testing, numerical simulations were performed to verify the predictive capability of the material model and assess its applicability for further use. Numerical simulations were performed using a constitutive viscoplastic model calibrated against tensile test data, to verify the predictive capability and applicability of the material model.

**Keywords:** PA6 GF; Grivory; tensile testing; relaxation; viscoplasticity.

## 1 Introduction

Structural polymers are widely used in engineering applications where components are exposed to various loading conditions, including time-dependent effects. Understanding the mechanical behavior of these materials is crucial for ensuring their long-term performance and durability [4]. This study focuses on polyamide 6 reinforced with glass fibers (PA6 GF, Grivory), a material commonly employed in load-bearing structures. Due to its semi-crystalline nature, PA6 GF exhibits both strain-rate-dependent and time-dependent deformation behavior, thus making experimental characterization essential for accurate material modeling [5].

Polymeric materials, especially polyamide 6 reinforced with glass fibers (PA6 GF), are widely used in structural applications due to their high strength and stiffness. However, their mechanical behavior is influenced by time-dependent phenomena such as creep and stress relaxation, which can affect the long-term performance of components [8]. Ahmadifar et al. (2023) [1] demonstrated that the reinforcement of PA6 with glass fibers significantly impacts its tensile and fatigue properties, which is critical for applications requiring long service life. Dencheva et al. (2011) [2] further showed that the shape and orientation of the reinforcement play a key role in the mechanical behavior of PA6 composites. Moreover, studies on the creep behavior of PA6 GF materials [3] highlight the importance of aging on changes in mechanical properties, which is essential for predicting long-term reliability (Creep behavior of glass-fiber-reinforced nylon 6 products).

This work investigates the tensile and relaxation behavior of PA6 GF through an extensive experimental program. Tensile tests at multiple strain rates provide insight into the material's viscoplastic response, while relaxation tests performed using a stepwise deformation method allow for the identification of critical strain levels where time-dependent effects become significant. These findings provide valuable data for predictive modeling, numerical simulations, and engineering applications.

## 2 Experimental Methodology

### 2.1 Material and Specimen Design

The material investigated is Grivory GV-4H black, a 40% glass fiber-reinforced thermoplastic based on a semi-crystalline polyamide (PA6) blended with a partially aromatic copolyimide. This combination enhances mechanical strength, stiffness, and thermal stability, making it well-suited for structural applications exposed to demanding loading conditions [13].

The addition of short glass fibers significantly improves the material's modulus and tensile strength, while also affecting its fracture behavior and anisotropy, depending on the fiber orientation induced by processing. Compared to unreinforced PA6, the presence of glass fibers reduces moisture absorption, while enhancing dimensional stability and maintaining mechanical properties, even in humid environments. Additionally, Grivory GV-4H exhibits high heat resistance, allowing for use in applications where elevated temperatures and long-term mechanical loads are critical factors [13].

To ensure consistent geometry and repeatability, specimens were fabricated following standardized guidelines. A small-sized specimen type 1BA was selected, conforming to ČSN EN ISO 527-2, with a cross-section of 4 × 5 mm [6]. Fiber orientation is influenced by the injection molding process, which can lead to direction-dependent mechanical properties.

The geometry of the tested specimen is illustrated in (Fig 1a), and (Fig 1b) shows a specimen clamped in the testing machine with an extensometer attached.

### 2.2 Tensile Testing Procedure

Tensile tests were performed on a universal testing machine equipped with a linear actuator, which controlled the crosshead displacement at speeds of 1, 10, and 100 mm/min. The tests were conducted at room temperature using a contact extensometer Epsilon for precise strain measurement. The resulting stress-strain curves were recorded and processed to evaluate the material's response to different loading conditions [6].

### 2.3 Relaxation Testing Procedure

Relaxation tests were conducted using an accelerated stepwise loading method. The specimens were gradually loaded to predefined strain levels and held at a constant strain for 300 seconds at each step. The primary relaxation tests were performed at strain levels of 0.5%, 1%, 1.5%, and 2% to analyze time-dependent stress changes at higher strain levels.

To investigate the viscous effects in the elastic region, additional relaxation tests were carried out at strain levels ranging from 0.1% to 0.5%, allowing for the assessment of the material's response in the low-strain range.

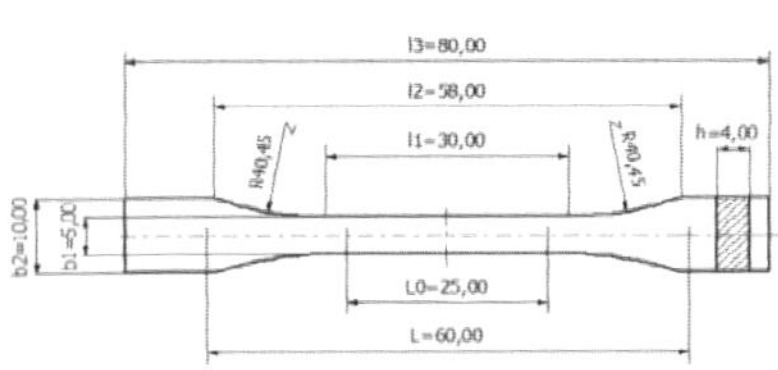


a) Geometry of the 1BA tensile specimen according to ČSN EN ISO 527-2.

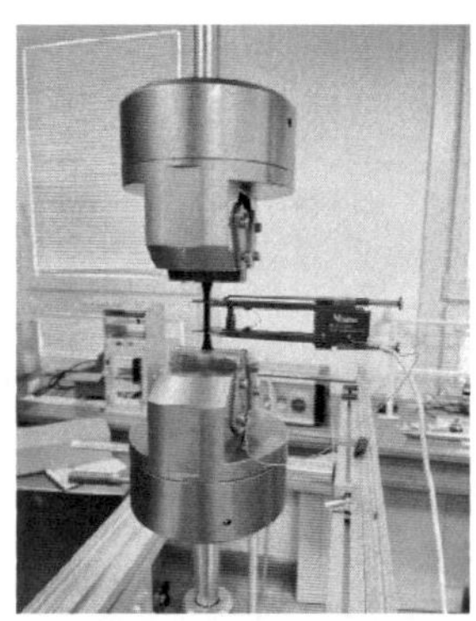

b) Specimen clamped in the testing machine with an attached extensometer.

Fig. 1: Specimen geometry and experimental setup.

# 3 Results and Discussion

The experimental results provide insight into the mechanical response of PA6 GF (Grivory) under various loading conditions. The data collected from tensile tests illustrate the material's strain-rate dependency, while relaxation tests characterize its time-dependent behavior. This section presents an overview of the observed trends and provides an interpretation of the findings based on experimental data [9].

## 3.1 Tensile Test Results

The stress-strain curves obtained at different strain rates show variations in ultimate strength and ductility, while Young's modulus remained nearly constant. Engineering stress vs. engineering strain curves for each strain rate are shown in (Fig 2a). The increase in ultimate strength with higher loading speeds confirms the strain-rate sensitivity of PA6 GF.

Young's modulus was determined using linear regression in the elastic (linear) region of the stress-strain curve, as shown in Fig 2b.

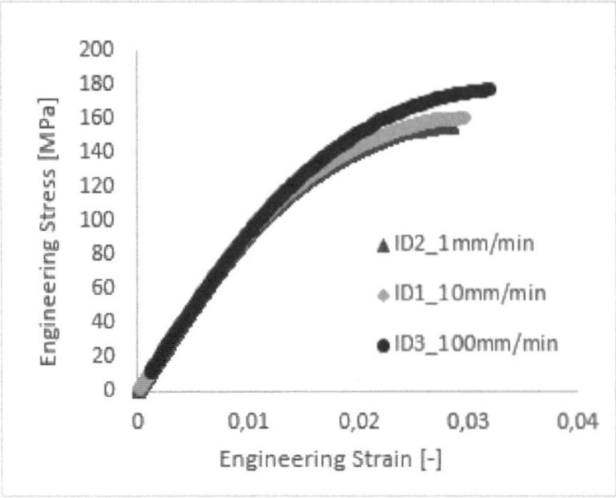


a) Engineering stress vs. engineering strain curves for tensile tests at strain rates of 1, 10, and 100 mm/min.

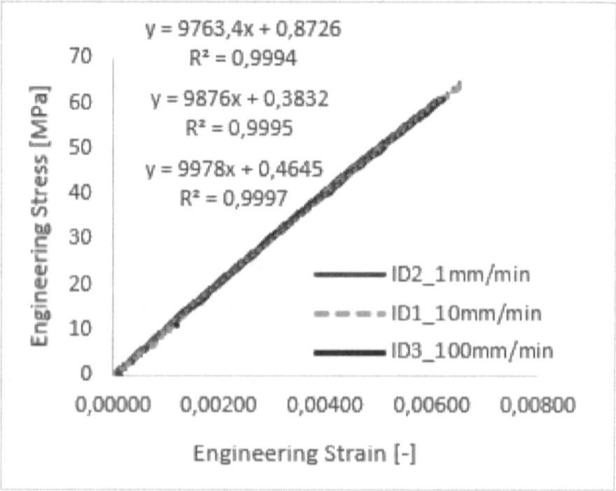


b) Young's modulus determined using linear regression in the elastic (linear) region of the engineering stress-strain curve.

Fig. 2: Tensile test results and Young's modulus evaluation.

Tab. 1: Mechanical properties of PA6 GF obtained from tensile tests.

| Crosshead speed [mm/min] | Young's modulus [MPa] | Ultimate tensile strength, Rm [MPa] | Strain at break [%] |
|---|---|---|---|
| 1 | 9978 | 155 | 2.9 |
| 10 | 9876 | 160 | 3 |
| 100 | 9763 | 176 | 3.19 |

## 3.2 Relaxation Test Results

The relaxation tests conducted at selected strain levels provide data on the evolution of stress over time. The results were analyzed to identify the strain threshold at which time-dependent effects become significant.

Within the higher strain range (0.5%–2%), a noticeable stress relaxation was observed, confirming the material's viscoplastic response. In contrast, at lower strain levels (0.1%–0.5%), the stress decay was minimal, particularly below 0.3% strain, where relaxation effects were negligible. This confirms that PA6 GF exhibits predominantly elastic behavior up to this threshold [12].

Graphs displaying relaxation curves for higher strain levels are presented in (Fig 3), and (Fig 4) illustrates the results for lower strain levels.

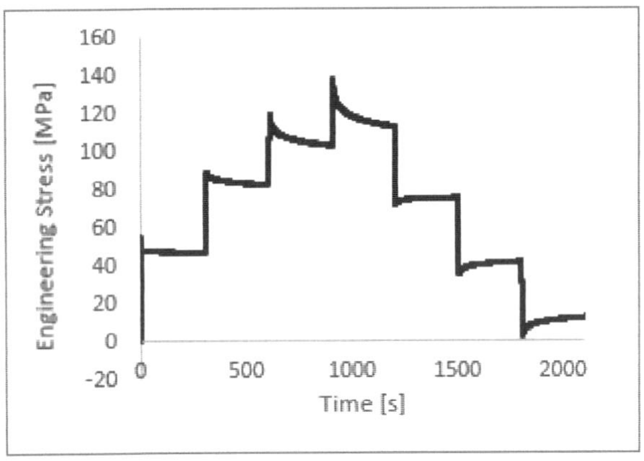


a) Stress relaxation curves showing the evolution of stress over time for selected strain levels.

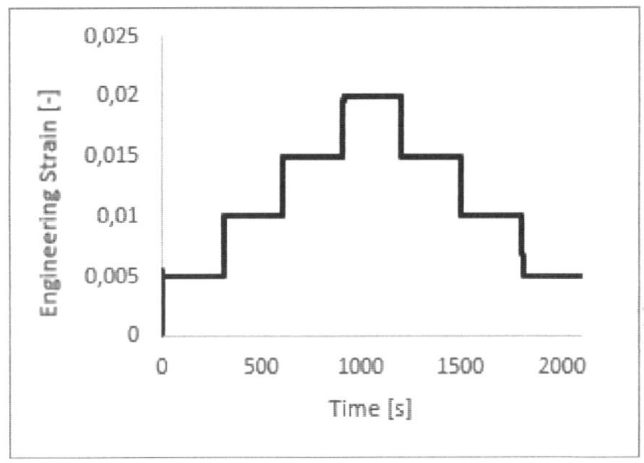


b) Applied strain over time, illustrating the stepwise deformation approach used in the relaxation tests.

Fig. 3: Relaxation test results for higher strain levels.

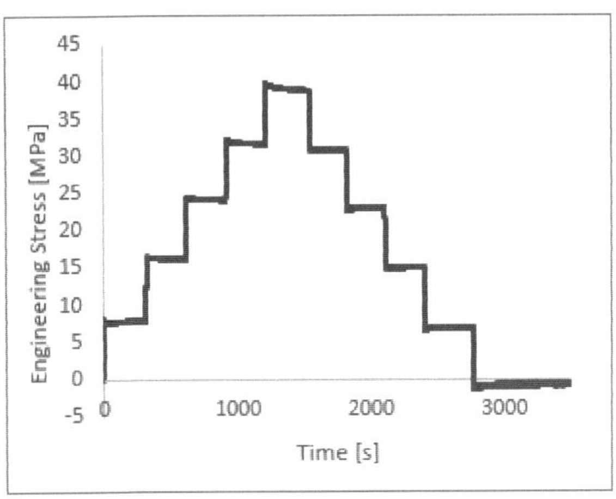


a) Stress relaxation curves showing the evolution of stress over time for strain lower levels.

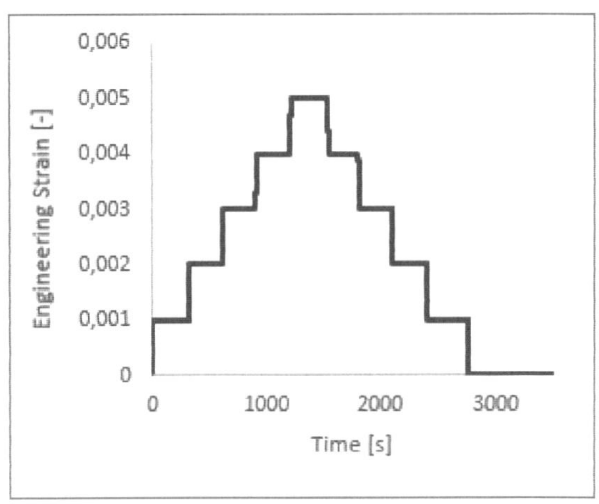


b) Applied strain over time, illustrating the stepwise deformation approach used in the relaxation tests.

Fig. 4: Relaxation test results for lower strain levels.

### 3.3 Summary of Experimental Findings

The experimental results demonstrated that Young's modulus remained nearly constant across all tested crosshead speeds, indicating that the material's elastic response is not significantly rate-dependent. The ultimate tensile strength increased with higher loading speeds and confirmed the viscoplastic nature of PA6 GF.

Relaxation tests revealed that significant stress relaxation occurred at strain levels above 0.5%, confirming time-dependent viscoplastic behavior. In contrast, minimal relaxation was observed below 0.3% strain, indicating a predominantly elastic response. The transition between elastic and viscoplastic behavior was identified at approximately 0.3% strain, providing a threshold for further material modeling.

## 4 Numerical Modeling

Numerical simulations were conducted to model the mechanical response of PA6 GF (Grivory) under tensile and relaxation conditions. The simulations were performed in ANSYS 2023 R2, where a viscoplastic material model combining Perzyna's viscoplasticity with Chaboche's kinematic hardening is implemented. The governing constitutive equation describes the strain-rate-dependent and time-dependent response of the material.

$$\sigma = \sigma_y + \left(\frac{C_1}{\gamma_1}\cdot(1-e^{-\gamma_1\cdot\varepsilon_p}) + \frac{C_2}{\gamma_2}\cdot(1-e^{-\gamma_2\cdot\varepsilon_p}) + \frac{C_3}{\gamma_3}\cdot(1-e^{-\gamma_3\cdot\varepsilon_p}) + C_4\cdot\boldsymbol{\varepsilon_p}\right) + K\cdot\dot{\boldsymbol{\varepsilon_p}}^{m} \quad (1)$$

### 4.1 Model Calibration

The model was initially calibrated using data from tensile tests, ensuring an accurate representation of the stress-strain behavior. After obtaining relaxation test results, the yield stress was further adjusted to correspond with the experimentally observed transition between elastic and viscoplastic behavior. This modification ensured that the model captured the elastic threshold identified at approximately 0.3% strain. The material parameters used in the numerical model are summarized in Table 2.

Tab. 2: Calibrated material parameters used in the numerical model.

| Constant | $\sigma_y$ | $C_1$ | $\gamma_1$ | $C_2$ | $\gamma_2$ | $C_3$ | $\gamma_3$ | $K$ | $m$ |
|---|---|---|---|---|---|---|---|---|---|
| Value | 30 | 46910 | 2115 | 18963 | 263 | 3.92e-15 | 0 | 93 | 7,022e-05 |

### 4.2 Simulated Tests and Comparison with Experiment

The tensile and relaxation tests were simulated on a unit-sized specimen, maintaining consistency with experimental conditions. The simulated stress-strain curves were compared with experimental data to evaluate the model's predictive capability. Additionally, relaxation simulations were performed to verify the time-dependent stress decay behavior.

Fig 4 and 5, 6 display the comparison between experimental and simulated results for tensile and relaxation tests, respectively. The model accurately predicts the strain-rate sensitivity observed in tensile tests and captures the stress relaxation trends, although deviations exist due to the simplified material representation.

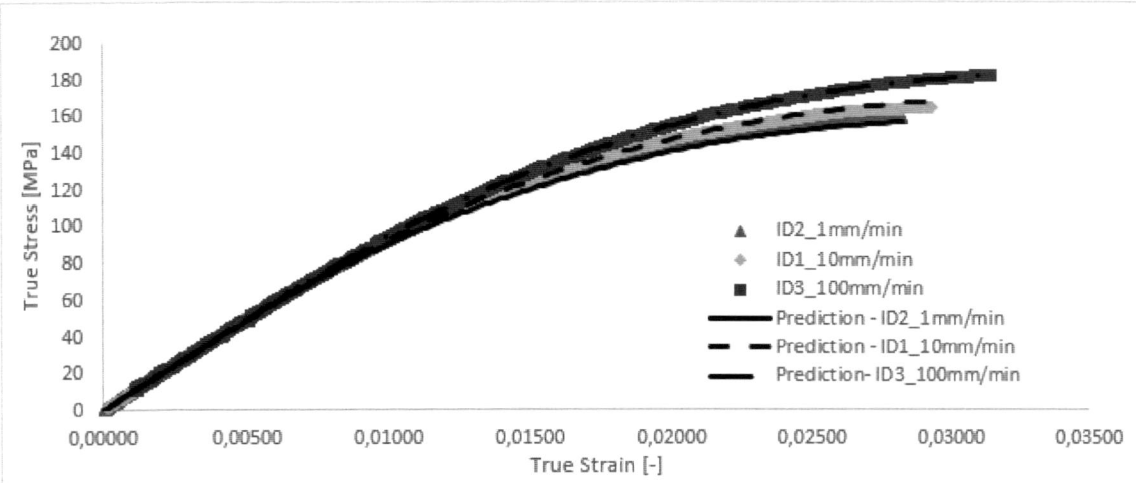


Fig. 4: Comparison of experimental and simulated tensile test results.

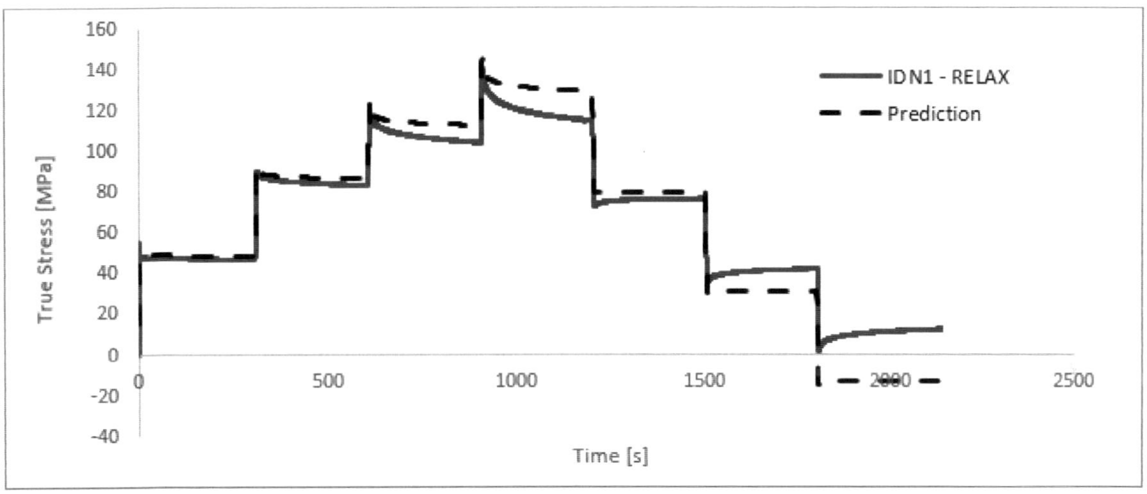


Fig. 5: Comparison of experimental and simulated test results over time for 0.5 – 2 % strain levels.

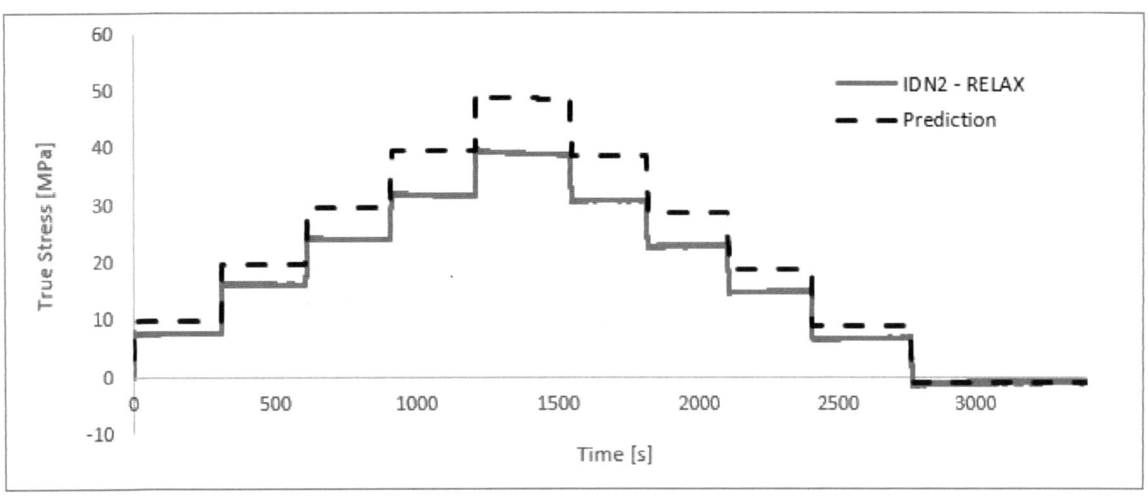


Fig. 6: Comparison of experimental and simulated test results over time for 0.1 – 0.5 % strain levels.

## 5 Conclusion

The experimental and numerical investigation provided a detailed characterization of the mechanical behavior of PA6 GF (Grivory) under tensile and relaxation loading regime.

Tensile tests confirmed that Young's modulus remains nearly constant. The ultimate tensile strength increased with increasing loading rate, confirming the viscoplastic nature of the material. Relaxation tests revealed significant stress relaxation at strain levels above 0.5%. Minimal relaxation was observed below approximately 0.3% strain, indicating predominantly elastic behavior in this region. The transition between elastic and viscoplastic response was identified at approximately 0.3% strain, providing a crucial reference point for further material modeling.

Numerical simulations of conducted tensile tests at various loading rates showed excellent agreement with experimental results, confirming the model's capability to capture the rate-dependent response. Simulations of relaxation tests, however, exhibited discrepancies. At lower strain levels, the predicted stresses were overestimated. At higher strain levels, the predicted stress relaxation was underestimated. Experimental data clearly showed an immediate reduction in plastic deformation during unloading, whereas the current numerical model predicted this behavior only at very low stresses (approximately twice the yield stress). This effect leads to significant inaccuracies during unloading simulations, highlighting the need to modify the current constitutive model by incorporating an isotropic viscoplastic component or other features to accurately capture the immediate decrease of plastic deformation [10].

The findings presented will serve as a basis for further improvement of numerical models and for planned future research activities involving creep and fatigue testing [4,7,11].

## Acknowledgement

This work was supported by Specific Research „Computational and Experimental Modeling in Applied Mechanics and Biomechanics "(SP2025/048).

## References

[1] M. Ahmadifar, K. Benfriha, M. Shirinbayan, Thermal, Tensile and Fatigue Behaviors of the PA6, Short Carbon Fiber-Reinforced PA6, and Continuous Glass Fiber-Reinforced PA6 Materials in Fused Filament Fabrication (FFF), Polymers 15[1] (2023) 507, https://doi.org/10.3390/polym15030507.

[2] N.V. Dencheva, et al., Mechanical properties of polyamide 6 reinforced microfibrilar composites, Polymer Composites 32 [1] (2011) 407–417, https://doi.org/10.1002/pc.21056.

[3] H. Tohmyoh, Y. Ito, et al., Creep behavior of glass-fiber-reinforced nylon 6 products, Journal of Applied Polymer Science 124 [2] (2012) 4213–4221, https://doi.org/10.1002/app.35393.F. Author, Title of Book, City of Publication, Name of Publisher, 2002.

[4] P.N. Mallick, Y. Zhou, Effect of mean stress on the stress-controlled fatigue of a short E-glass fiber reinforced polyamide-6,6, International Journal of Fatigue 26[3] (2004) 941–946, https://doi.org/10.1016/j.ijfatigue.2004.02.003.

[5] O.-L. Petrașcu, A. Pascu, Comparative study of polyamide 6 (PA6) and polyamide 6 reinforced with 30% of glass fiber (PA6GF30), Materials Today: Proceedings 93[3] (2023), https://doi.org/10.1016/j.matpr.2023.04.112.

---

[1] Journal volume.

[2] Journal volume.

[3] Journal volume.

[6] Plastics – Determination of tensile properties – Part 1: General principles, ČSN EN ISO 527-1:2019, Czech Office for Standards, Metrology and Testing.

[7] Zadravec, M., et al., S. Fatigue Behaviour of PA66 GF30 at Different Temperatures. Polymers 2025, 17, 42. https://doi.org/10.3390/polym17010042.

[8] Vikram, K., Pramanik, S., & Bhaumik, S. Thermal kinetic, mechanical and wear properties of glass fiber reinforced polyamide 6 thermoplastic composites. Journal of Applied Polymer Science, 2024, e56119. https://doi.org/10.1002/app.56119.

[9] Wang, Z., Zhou, Y., & Mallick, P.K. Effects of Temperature and Strain Rate on the Tensile Behavior of Short Fiber Reinforced Polyamide-6. Polymer Composites, 23(5), 858–870 (2002).

[10] Soós, E., & Goda, T. Mechanical behavior of glass fiber-reinforced bosses: experiments and FE simulations. Periodica Polytechnica Mechanical Engineering, 55(1), 3–14. https://doi.org/10.3311/pp.me.2011-1.01

[11] Regrain, C., Laiarinandrasana, L., & Toillon, S. Experimental and numerical study of creep and creep rupture behavior of PA6. Engineering Fracture Mechanics, 76(18), 2871–2885 (2009). https://doi.org/10.1016/j.engfracmech.2009.04.016

[12] Li, B., Liao, et al., Investigation on the tensile rheological behavior of PA6 film based on fractional order model. Materials Research Express, 10(10), 105302 (2023). https://doi.org/10.1088/2053-1591/acfd11

[13] EMS-GRIVORY, Product Datasheet Grivory GV-4H black 9915. EMS-CHEMIE AG, 2022.

# Effect of Waste Glass Surface Modification on the Mechanical Properties of Cement Composites

Z. Prošek [1,a], A. Palička [1], O. Babčenko [2]

[1] *Department of Mechanics, Faculty of Civil Engineering, CTU – Czech Technical University in Prague, Thákurova 7, 166 29 Prague, Czech Republic*
[2] *Institute of Physics, Academy of Sciences of the Czech Republic, Cukrovarnická 10, 162 00 Prague, Czech Republic*
[a] *zdenek.prosek@fsv.cvut.cz*

**Abstract:** This paper focuses on the study of photovoltaic waste glass used as a partial replacement for cement and its effect on the mechanical properties of cement composites. The surface of the glass powder was modified through treatments in oxygen atmospheres. The evaluated properties included compressive strength and flexural strength at the macro level. Testing was performed on specimens measuring 100 × 20 × 20 mm after 28 days and 3 months of curing. The results show that appropriate surface modification of waste glass can significantly influence the mechanical performance of the resulting composite.

**Keywords:** waste glass, surface modification, cement composites, compressive strength, flexural strength.

## 1 Introduction

The growing global demand for sustainable construction materials has driven research into secondary raw materials that can partially or fully substitute Portland cement in cementitious composites. Among these, waste glass has attracted attention due to its high silica content and its inherent pozzolanic potential, especially when finely ground. The recycling of waste glass aligns with the principles of circular economy and contributes to reducing the carbon footprint associated with cement production [1].

However, one of the main limitations of using glass powder in cementitious systems is the smooth, hydrophobic nature of its surface, which leads to poor bonding within the interfacial transition zone (ITZ). This weak ITZ results in a compromised microstructure, reducing both strength and durability [2].

Recent research has demonstrated that physical or chemical modification of the glass surface can increase its reactivity and compatibility with the cement matrix. Among various surface treatments, plasma treatment in an oxygen atmosphere has proven effective in increasing surface energy, introducing oxygen-containing functional groups, and thus improving the adhesion between the glass particles and the surrounding hydration products [3].

While most studies to date focus on packaging waste glass, which is widely available and chemically well-characterized, the recycling of end-of-life photovoltaic (PV) panels presents a new challenge and opportunity. The decommissioning of PV panels is expected to surge in the coming years, generating significant volumes of high-purity glass. However, the glass in PV panels often includes additional surface coatings, antireflective layers, or embedded conductive films, which may influence its reactivity and compatibility with cementitious systems [4,5].

Understanding these differences is essential for the development of effective recycling strategies and for promoting circular economy principles in the construction industry. Further research is required to evaluate whether PV waste glass, when properly treated, can match or even surpass the performance of conventional waste glass in cement composites, contributing to resource efficiency and environmental sustainability.

## 2 Materials and Samples

All test mixtures were based on Portland cement CEM I 42.5R. Waste glass powder with particle sizes below 75 μm was used to replace 10 wt.% of cement. Two types of glass were used: (i) post-consumer packaging glass, predominantly soda-lime composition (PG), and (ii) photovoltaic glass extracted from dismantled solar panels (PV).

Both glass types were tested in untreated form and after surface modification using low-pressure oxygen plasma. The plasma treatment was conducted in a chamber at 15 Pa pressure and 600 W input power for 5 minutes. These parameters were selected based on previous studies to optimize the balance between surface activation and potential material degradation.

Prismatic samples with dimensions of 100 × 20 × 20 mm were cast for all mixture variants. The specimens were demolded after 24 hours and subsequently cured in water at 22 ± 1 °C for two distinct periods: 28 days and 90 days. The composition of individual mixes and the workability of fresh mixtures can be seen in Tab 1.

Tab. 1: Composition of the testing samples.

| Mixture | Cement [wt.%] | Waste glass [wt. %] | Glass Type | Surface Treatment |
|---|---|---|---|---|
| Ref. | 100 | 0 | – | – |
| PV_Ref. | 90 | 10 | Photovoltaic | None |
| PV_O | 90 | 10 | Photovoltaic | $O_2$ Plasma |
| PG_Ref. | 90 | 10 | Packaging | None |
| PG_O | 90 | 10 | Packaging | $O_2$ Plasma |

## 3 Experimental methods

Flexural strength was determined using a displacement-controlled three-point bending test. The setup had a support span of 80 mm, tailored to the specimen size of 100 × 20 × 20 mm. The loading rate was set to a constant crosshead displacement of 0.1 mm/s to ensure quasi-static conditions and minimize inertial effects. The force-displacement data were recorded continuously, and the peak load at fracture was used to compute flexural strength based on the beam theory for prismatic specimens.

Following the flexural test, the fractured halves of each specimen were carefully aligned and tested in uniaxial compression to determine the compressive strength. The loading area corresponded to the effective cross-section of 20 × 20 mm. Compression testing was performed under displacement control at a rate of 0.3 mm/s. This method allowed for direct comparison of compressive behavior across all material variants while maintaining efficient specimen utilization.

All tests were carried out at ambient laboratory conditions using a calibrated universal testing machine with a load capacity suitable for the expected range of forces (up to ~100 kN). Testing adhered to procedures consistent with relevant EN and ISO standards, ensuring comparability with established literature.

## 4 Result s and Discussion

The mechanical performance of all cement composite mixtures was evaluated through flexural (Fig. 1) and compressive strength (Fig. 2) tests after 28 and 90 days of water curing. The obtained results clearly demonstrate the significant influence of both the type of waste glass and the applied surface treatment on the strength development of the composites.

At 28 days, the reference mixture without any glass addition exhibited the highest flexural strength of 11.63 MPa. Incorporation of untreated PV glass led to a reduction in flexural strength to 9.50 MPa, indicating a weakening of the ITZ due to the inert and smooth surface of the glass particles. Packaging glass without treatment performed slightly better, reaching 10.60 MPa. However, plasma treatment significantly improved the flexural performance of both glass types. PV glass treated in oxygen plasma reached 10.50 MPa at 28 days, nearly matching the untreated packaging glass. Plasma-treated packaging glass achieved 9.92 MPa, which, while slightly lower than its untreated counterpart, still shows stable behavior. After 90 days of curing, all mixtures demonstrated further strength development. The reference mixture increased to 12.50 MPa, while PV and packaging glass mixtures also improved. Notably, plasma-treated PV glass reached 11.50 MPa, and plasma-treated packaging glass 11.00 MPa, confirming the long-term beneficial effect of surface activation on ITZ integrity.

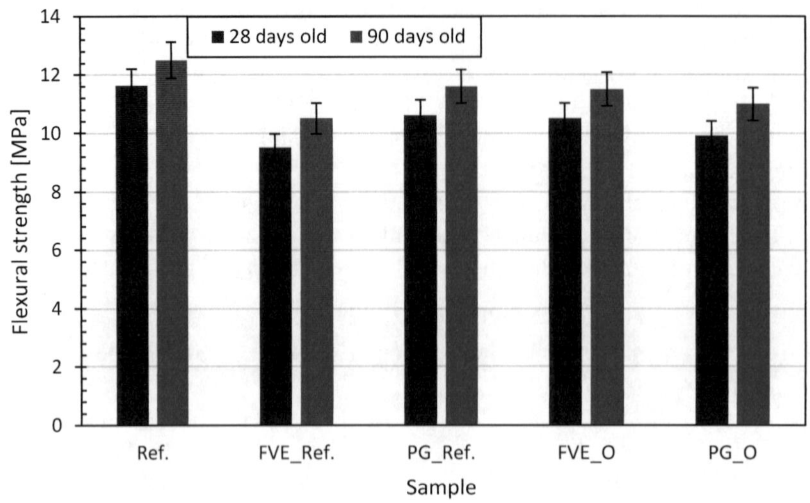


Fig. 1: Average value of flexural strength of cement composites after 28 and 90 days with standard deviation.

A similar trend was observed in compressive strength results. The reference composite achieved 79.87 MPa after 28 days and 101.49 MPa after 90 days. Replacement of cement with untreated PV glass reduced compressive strength to 62.00 MPa at 28 days and 72.00 MPa at 90 days. This confirms that without surface activation, the glass impairs matrix continuity and delays strength gain. Plasma treatment significantly mitigated these adverse effects. The 28-day compressive strength of plasma-treated PV glass rose to 65.00 MPa, and by 90 days it reached 90.00 MPa, representing a 25% improvement over the untreated variant. Similarly, packaging glass samples showed clear improvement. Plasma-treated samples achieved 65.21 MPa at 28 days and 97.70 MPa at 90 days, compared to 68.37 MPa and 71.39 MPa for untreated packaging glass.

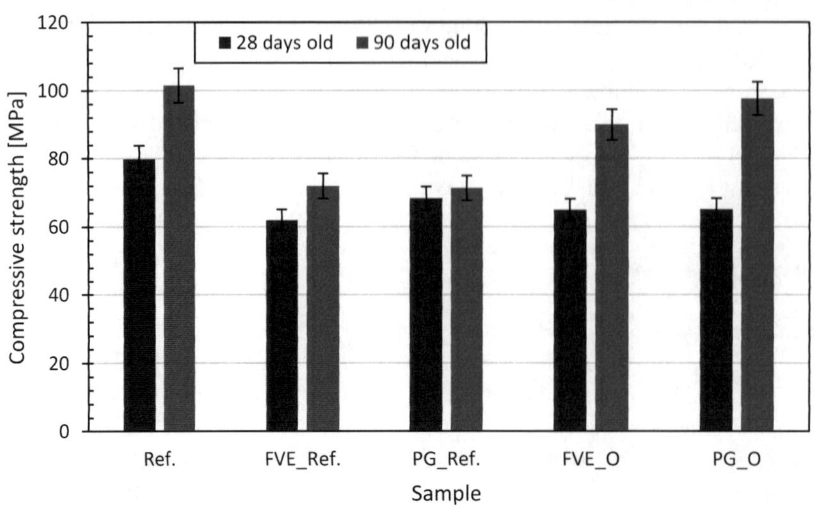


Fig. 2: Average value of compressive strength of cement composites after 28 and 90 days with standard deviation.

The results confirm the general observation that the inclusion of untreated waste glass reduces mechanical performance due to its poor reactivity and weak bonding with the cementitious matrix. However, oxygen plasma treatment significantly improves both flexural and compressive strength, particularly over longer curing times. Plasma activation likely enhances surface roughness and introduces polar functional groups that increase the pozzolanic activity of glass particles. This, in turn, promotes better integration with the hydration products, densification of the ITZ, and delayed but more robust strength development. The findings demonstrate that properly treated photovoltaic waste glass can match or even exceed the performance of conventional packaging glass, making it a viable candidate for use as a supplementary cementitious material (SCM) in sustainable construction.

# 5 Conclusion

This study provides experimental evidence that oxygen plasma surface treatment significantly improves the mechanical properties of cement composites containing waste glass, including less conventional sources such as PV panel glass. The most pronounced improvement was observed in long-term compressive strength, indicating that the plasma-induced surface changes continue to positively influence the hydration process beyond the initial curing period. From a practical perspective, these findings support the viability of using finely ground PV waste glass as a supplementary cementitious material (SCM), provided it undergoes appropriate pretreatment. This opens new avenues for managing PV waste streams in the construction industry and contributes to sustainable development goals through reduced use of primary raw materials and lower carbon emissions.

Further research will focus on a more detailed microstructural characterization of the interfacial transition zone using SEM, FIB-SEM, and nanoindentation techniques. The goal is to quantify the effects of plasma treatment on bond strength and to model the long-term durability of such composites in aggressive environments. In addition, the influence of treatment parameters (e.g., duration, pressure, plasma gas composition) on mechanical and chemical performance will be systematically explored.

# Acknowledgement

This paper was financially supported by the GA CR research a project under the number GA23-05500S. The authors also thank Recifa a.s.

# References

[1] A. M. Neville, Properties of Concrete, 4th Edition, Longman, London, 1995.

[2] Y. Ke, et al., Identification of microstructural characteristics in lightweight aggregate concretes by micromechanical modelling including the interfacial transition zone (ITZ), Cement and Concrete Research, 40 (2010), pp. 1590–1600. https://doi.org/10.1016/j.cemconres.2010.07.001.

[3] Z. Prošek, V. Nežerka, P. Tesárek, Enhancing cementitious pastes with waste marble sludge, Construction and Building Materials, 255 (2020), 119372. https://doi.org/10.1016/j.conbuildmat.2020.119372.

[5] Z. Prošek, A. Palička, P. Tesárek, Characterisation of the Interfacial Transition Zone Between the Cement Matrix and Glass Powder Particles, in: Proceedings of the 62nd International Conference on Experimental Stress Analysis (EAN 2024), Book of Extended Abstracts, University of West Bohemia in Pilsen, 2024, ISBN 978-802611225-9, Bozi Dar, Czech Republic, June 4–6, 2024.

[6] Z. Prošek, et al., Mechanical Properties of Cement Pastes Containing Plasma-Treated Silicon-Based Materials, Acta Polytechnica CTU Proceedings, vol. 47, 2024, pp. 108–112. https://doi.org/10.14311/APP.2024.47.0108.

# Development and Experimental Validation of Sensors for Structural Health Monitoring of Composite Structures

M. Růžička[1,*], N. Schmidová[1], K. Doubrava[1], D. Blaha[1], D. I. Exner[1], T. Ridoško[2]

[1] *Division of Strength and Elasticity, Faculty of Mechanical Engineering, Czech Technical University in Prague; Technická 4, 166 07, Prague, Czech Republic.*

[2] *OHB Czechspace s.r.o., Purkynova 648/125, 612 00 Brno-Medlanky, Czech Republic.*

[*] *milan.ruzicka@fs.cvut.cz*

**Abstract:** This paper presents the development and experimental validation of a hybrid sensor system for Structural Health Monitoring (SHM) of composite structures. The sensor combines Distributed Fiber Optic Sensing (DFOS) and Carbon Fiber Sensors (CFS) with both optical and piezoresistive properties. It can be integrated into the composite laminate or applied to its surface. Composite tube specimens were tested under bending and axial compression, both with and without the SHM system, to assess sensor impact on structural strength. Impact tests and ultrasonic testing were used to evaluate damage detection capability. The developed solution shows potential for long-term SHM of composites in the aerospace, automotive, and energy sectors.

**Keywords:** Structural Health Monitoring; Distributed Fiber Optics Sensors; Carbon Fiber Sensors; Composite Tube; Impact Testing; Nondestructive Testing.

## 1 Introduction

Structural Health Monitoring (SHM) is a field of research and technology focused on monitoring the condition of structures to enable early detection of damage and material degradation. SHM employs various methods and sensors for continuous or periodic analysis of structural changes [1]. Key approaches in SHM include fiber optic sensors: the Fiber Bragg sensor (FBG) or the Distributed Fiber Optic Sensing sensor (DFOS), which offer high sensitivity to deformations and are suitable for long-term monitoring of large areas, allowing stress and strain measurement in the area of the Bragg grid or along the entire fiber length. Piezoelectric sensors are used to detect mechanical vibrations and ultrasonic signals associated with structural damage, operating by generating an electric charge under mechanical stress. Resistive sensors based on carbon fibers utilize the piezoresistive properties of carbon microfibers, enabling their integration into composite structures and providing insights into mechanical stress changes and potential damage. In addition, strain gauges (SG) can be used for both types of sensor verification. An SHM system integrated into a composite structure, specifically composite wound tubes, should meet several key requirements: easy technological implementation in mass production, seamless integration of sensors, including connectors, to ensure straightforward connectivity with measuring equipment. In addition, the sensor system should have minimal impact on the structural properties of the composite component, reducing its ultimate strength and fracture limits as little as possible.

## 2 Development and integration of Combined Sensor

### 2.1 Development of Combined Sensor

For better manipulation and sensor placement during its integration into the composite tube, the combined sensor was developed and placed on a fiberglass plate. A glass-fiber prepreg layer served as an envelope for the two types of sensors: CFS manufactured from Toray T300 carbon fiber and DFOS sensor with polyimide coating. The sensor lengths could be adapted to the part; in this case the lengths of the sensors were 100 mm and

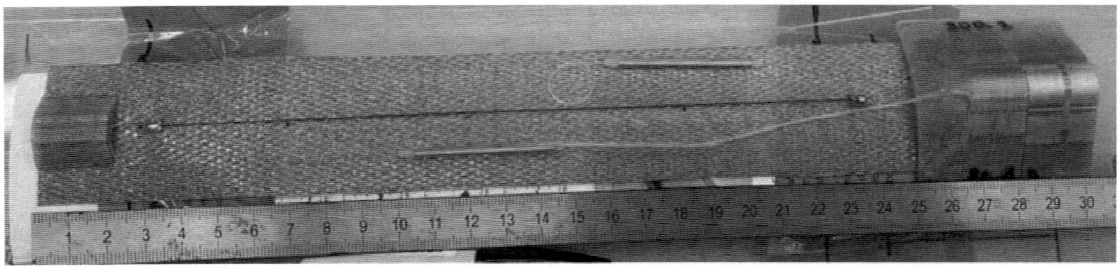


Fig. 1: Combined sensor design.

300 mm. For CFS connections, a proprietary process for galvanizing soldered fiber ends was developed. The sensors were made with plug-and-play connections for safety and easy to handle reasons. Together with this concept, the housing of the sensor connections was designed. The dimensions and shape of the housing were optimized according to the manufacturing process to achieve ±45° coverage with the last layer. The connectors in the housing were filled with a silicone-based cover to protect the connectors against matrix leakage. All materials used for the combined sensor, including the sensor housing and its connectors, were resistant to thermal cycling. The final combined sensor is shown in Fig. 1.

### 2.2 Calibration of the CFS Strain Sensitivity

The accuracy of the strain measurement with using of Carbon fiber sensors (CFS) with nickel-coated ends depends on determining the longitudinal and transverse sensitivities of the sensor. Although previous studies, such as [2], have investigated CFS sensitivity, reported values vary widely, making it difficult to select a definitive parameter. To address this, an experimental measurement was conducted [3]. Sixteen sensors were installed on a flat steel bar, including 12, ex-PAN T300B 3K, CFS of two different lengths to assess the influence of sensor length on sensitivity, and four strain gauges for reference. The specimen underwent a four-point bending test, ensuring a constant uniaxial tensile or compressive stress at the sensor locations. The total sensitivities $k$ and longitudinal $k_l$ and transverse $k_t$ were calculated by an iterative process using the following equations [2]:

$$k = k_l + k_t \frac{\varepsilon_y}{\varepsilon_x}, \quad \Delta R_{CFS_T}/R_{350} = k_l \varepsilon_x + k_t \varepsilon_y \tag{1}$$

$$(\Delta R/R_{CFS_L})_l = k_l \varepsilon_x + k_t \varepsilon_y, \quad (\Delta R/R_{CFS})_t = k_l \varepsilon_y + k_t \varepsilon_x \tag{2}$$

where $\Delta R$ is the resistance change, $R_{CFS_L}, \varepsilon_{CFS_L}$ represent the resistance and strain measured by the CFS sensors placed longitudinally, $R_{CFS_L}, \varepsilon_{CFS_T}$ denote the resistance and strain measured transversely, $\varepsilon_x$ and $\varepsilon_y$ are the principal longitudinal and transverse strains. The values of $\varepsilon_x$ and $\varepsilon_y$ were determined using strain gauges, taking into account the deformation sensitivities specified by the strain gauge manufacturer.

### 2.3 Integration of Combined Sensor in the Carbon Tube

For experimental validation of the functionality of the SHM system in future composite tubular frame structures, sensors were integrated during manufacturing into composite tubes made by filament winding using Toray T700K24 carbon fibers and the resin system Araldite LY 556 / Aradur 917 / Accelerator DY 070. This resin system was capable of withstanding the thermal cycling conditions applied to the specimens afterward. Specimen production took place at CompoTech.

The tubes had an inner diameter of 100.0 mm and the following basic layer configuration:

- 1. layer: T700, 0.26 mm thick, angle: 88.82°
- 2. layer: T700, 1.05 mm thick, angle: 0.00°
- 3. layer: T700, 0.88 mm thick, angle: ±45.59°
- Total thickness: 2.19 mm.

During sensor integration in this layout, the sensors were embedded inside the second axially wound layer (angle: 0°), resulting in a total wall thickness of 2.55 mm at the sensor location. The integration of the sensor on the axial layer is shown in Fig. 2.

Fig. 2: Integration of sensor.

# 3 Testing of the specimens with and without the SHM system

## 3.1 Configuration of the tests

### 3.1.1 Thermal testing

The prepared specimen were subjected to the thermal cycling to verify a thermal resistance without failure. The thermal cycling parameters were as follows:

- 50 cycles
- Lower limit: –50°C (–5/+0°C)
- Upper limit: +150°C (–0/+5°C)
- Temperature held until all sensors reported the target value, but not less than 5 minutes
- Temperature change rate: 10 ± 2°C/min
- Inert gas atmosphere

After thermal cycling, non-destructive ultrasound (NDT) was performed and no damage was detected.

### 3.1.2 Mechanical testing

The tubes were subsequently tested under two types of loading:

- Compression loading (tubes with a length of 700 mm equipped with steel end fittings to ensure even distribution of the compressive force)
- Four-point bending loading (tubes with a length of 1200 mm, with a support span of 1100 mm and a loading point span of 300 mm)

In the compression specimens, the single combined sensor with the length of 100 mm was integrated. In the bending specimens, two combined sensors were used: one 200 mm long on the top (compressed) side of the tube and the other 400 mm long on the bottom (tensile) side.

In addition to CFS and DFOS sensors, comparative continuous local strain measurements were also performed using HBM 1-LY41-6/350 strain gauges (manufacturer: Hottinger Brüel & Kjær GmbH, resistance: grid length: 6 mm, 350 Ω), connected in a one-quarter bridge configuration. For data acquisition, the HBM Quantum measurement system was used for both strain gauges and CFS, and the Luna ODiSI 6100 optical system was used for DFOS measurements.

Compression tests were performed using the Tiratest 2300 electromechanical test machine, and destructive compression tests were performed on the EU100 hydraulic test machine. Fig. 3 shows the compression test setup of a composite tube with steel end fittings. The bending tests were performed using the Heckert FZP100/1 electromechanical test machine.The configuration of the bending tests is shown in Fig. 4. To prevent local deformation, fabric straps were used for supports and loading points.

## 3.2 Example of the tests results

### 3.2.1 Compression tests

At the beginning of the test, the preload 100 μm/m was applied to the sample to secure the whole setup during the test. The specimens were loaded in compression (loading velocity about 1 kN/min) until a strain level of about 500 μm/m (on the axial reference strain gauge). Three cycles were performed at the same level.

Fig. 3: Compression test setup.

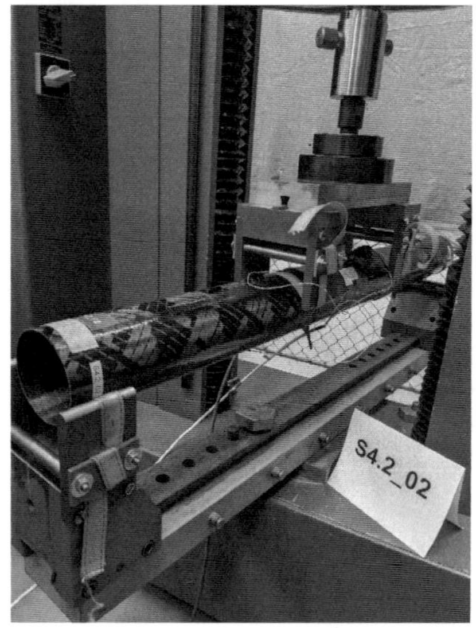


Fig. 4: Bending test setup.

The test was continued in three steps of cycles until a strain level of approximately 1000 µm/m was reached and then until a strain level of approximately 1500 µm/m was reached. The final loading was up to the maximum loading capacity of the Tiratest 2300 machine; see Fig. 5. A comparison of the averaged strain values of the signals from the strain gauges and the carbon fiber sensors dependent on load is shown in Fig. 6. The DFOS sensor could not be connected in this case due to a problem with the matrix that leaked into the connector housing.

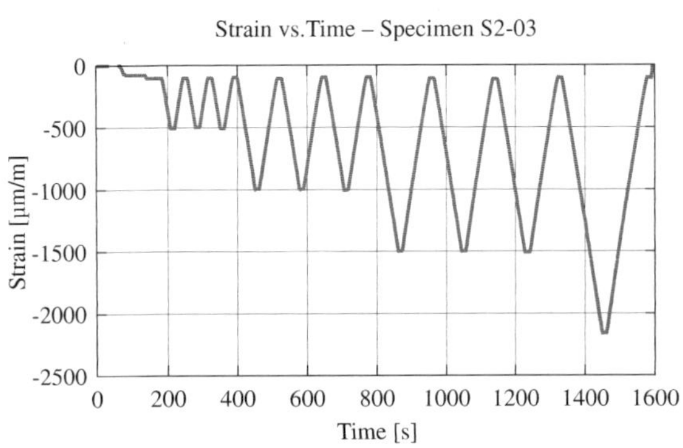


Fig. 5: Loading sequence of compression test.

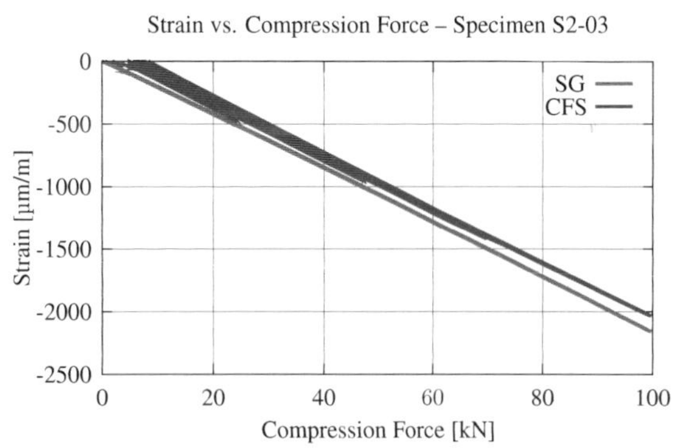


Fig. 6: Strain measured by SG and CFS.

Tab. 1: Table with compression tests summary.

| | Samples without SHM | | | Samples with SHM | | |
|---|---|---|---|---|---|---|
| Sample number | S1_01 | S1_02 | S1_03 | S2_01 | S2_02 | S3_03 |
| Rupture force [kN] | 328.9 | 337.9 | 333.3 | 140.5 | 125.7 | 115.3 |
| Average Value of rupture force [kN] | | 333.4 | | | 127.2 | |
| Diff. b/w laminate with and w/o SHM | | | -61.8 % | | | |

Destructive tests were performed in one load step on the EU100 test machine. The results of the obtained strengths are given in the Tab. 1. Comparison of the strength of the sample with and without the SHM sensor showed the limits of the proposed concept. The failure modes with and without SHM systems after the compressive destruction test are compared in Fig. 7.

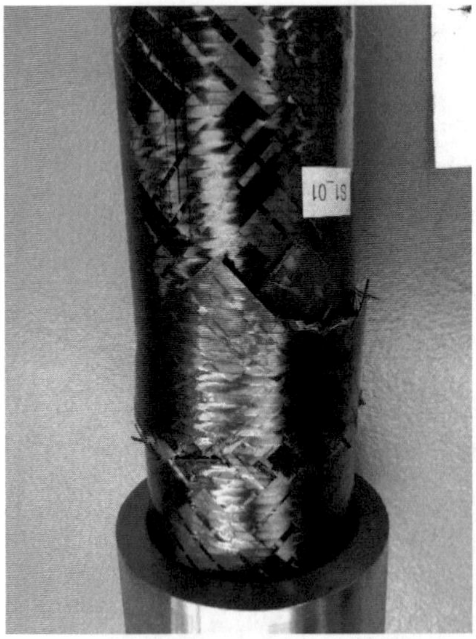

(a) Failure mode of specimen w/o SHM

(b) Failure mode of specimen with SHM

Fig. 7: Failure modes after compressive destruction test.

Tab. 2: Table with single bending test comparison.

| | Samples without SHM | Samples with SHM |
|---|---|---|
| Sample number | S3_01 | S4.1_02 |
| Min. strain – used as mach. contr. value [µm/m] | -1284 | -1269 |
| Applied force [kN] | 9.044 | 8.375 |
| Diff. b/w laminate with and w/o SHM | -7.7 % | |

### 3.2.2 Bending and Impact Tests

The 4BP tests of the specimens were performed to assess the detectability of damage caused by impact. The strain measured by the SGs installed on the surface of the specimen was the control load quantity. The loading was carried out similarly to that of the compressed specimens. The magnitude of the load levels was verified on the samples without SHM and a similar load was then applied to the specimens with SHM. A single comparison of the load levels for the samples with and without SHM is shown in Tab. 2.

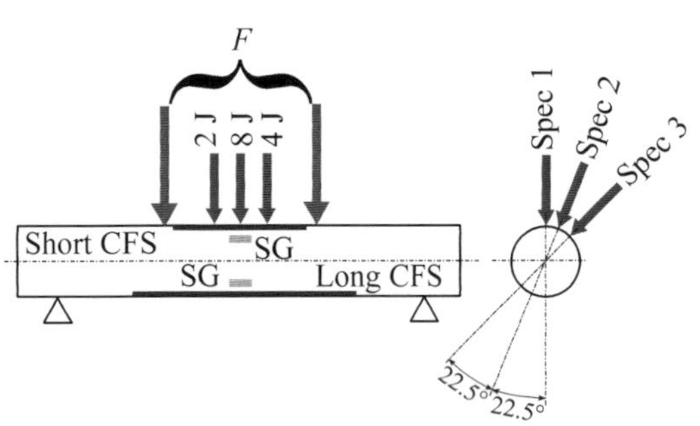


Fig. 8: Schema of 4PB and impact loading.

Fig. 9: Impact loading.

After determining the response of the sensors on the samples, an impact load was applied to the individual samples. The location and energy of the impacts are shown in Fig. 8. Each specimen was gradually affected by an increasing energy level (2 J, 4 J, 8 J). After each impact, a four-point bending test was performed to monitor the strain response of the sensor to see if it detected damage of the specimen. The distance of impact from the sensor installation location varied according to the diagram in Fig.8. The specimen sample during the impact test is shown in Fig. 9. Similarly to the compress specimen, the malfunction of the DFOS sensors was found there.

### 3.2.3 Improvement of the sensor housing

After problems with the functionality of the DFOS sensors (matrix leakage into the connector housing), a sample with a DFOS sensor with modified sensor integration technology was additionally manufactured. The functionality of the solution was verified by the 4PB loading using a similar procedure as in the previous case. The loading history from one point of the DFOS sensor can be seen in Fig. 10. According to the distributed sensing capability of the fiber optic sensors, the deformations along the length of the sensor can be seen in Fig. 11.

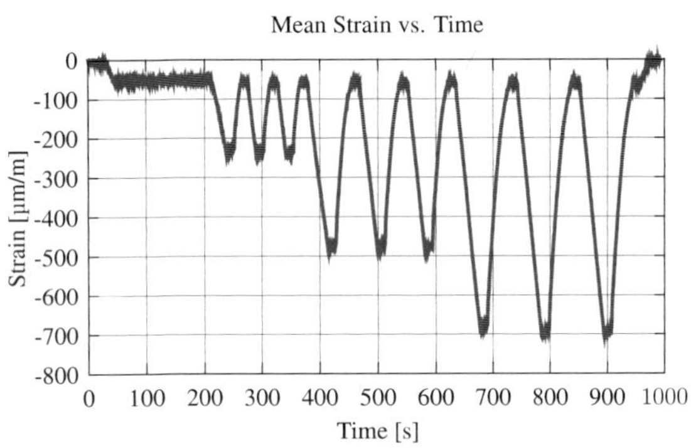


Fig. 10: Loading schema of 4PB test.

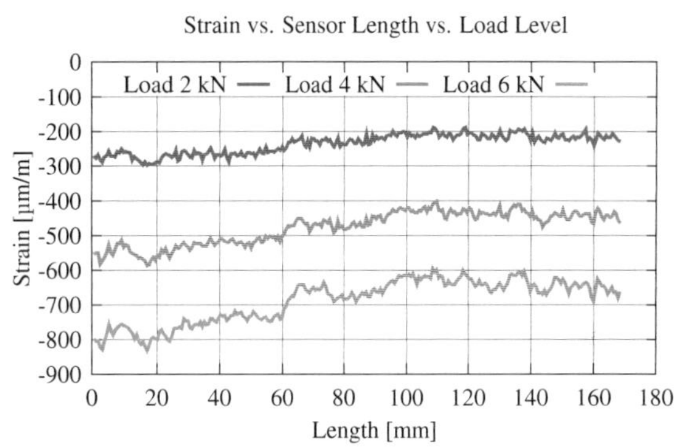


Fig. 11: Strain along the length of DFOS.

# 4 Conclusion

The sensitivity of CFS was investigated in previous studies [4], when the acquired experience was used to create a hybrid SHM sensor that combines DFOS and CFS. The novelty of this study lies in (i) the development of a combined sensor placed on a glass-fiber prepreg including a combined connector, (ii) the integration of this hybrid sensor into carbon composite tubes, and (iii) the comprehensive experimental validation of its performance under compression, bending, and impact loading.

# Acknowledgement

Financial support was provided for this study by the Grant Agency of the CTU in Prague; Grant no. SGS24/123/OHK2/3T/12, and project with the company Compotech Plus No 8302415B001. The project was also supported by FIRST! (Future Innovation Research in Space Transportation), and initiative of the ESA STS (Space Transportation) Future Launchers Preparatory Programme (FLPP).

# References

[1] C. R. Farrar, K. Worden, An introduction to structural health monitoring. Philosophical Transactions of the Royal Society A (2007), 365(1851), 303–315.

[2] A. Horoschenkoff, T. Muller, A. Kroell, On the characterization of the piezoresistivity of embedded carbon fibres, in ICCM International Conferences on Composite Materials, 2009.

[3] F. L. Jin, S. J. Park, Preparation and characterization of carbon fiber-reinforced thermosetting composites: A review, Carbon Letters (2015), 16(2), 67–77. https://doi.org/10.5714/CL.2015.16.2.067

[4] N. Schmidová et al., Bending Analysis of GFRP Composite Plate with a Hole Using Carbon Fiber Sensors, in proc. 62nd International Scientific Conference on Experimental Stress Analysis (EAN 2024), 4–6 June 2024, Boží Dar, Czech Republic.

# Mechanical and Thermal Properties Influenced Heat Treatment of Basalt Fibers-Vermiculite-Doped Geopolymer

M. Ryvolová [a], M. Mrózek, O. Vích, T. Bakalova

*Department of Material Science, Faculty of Mechanical Engineering, Technical University of Liberec, Studentská 2, 461 17 Liberec, Czech Republic*
*[a] martina.ryvolova@tul.cz*

**Abstract:** Composite materials represent a modern alternative to conventional building materials. One application of composite materials with the addition of vermiculite is the manufacture of safety boxes for electrical batteries, especially in environments exposed to high temperatures. Geopolymers in combination with vermiculite, a mineral whose crystals expand when heated and form light, porous, worm-like formations, open new possibilities in the field of fire prevention. The research describes the analysis of the potential of geopolymer material with combined reinforcement - vermiculite as an innovative solution for fire prevention, and basalt fibers to guarantee sufficient mechanical strength and durability in general. The newly designed composite materials provide sufficient protection against the spread of fire, which is confirmed by the achieved research results (thermal diffusivity and mechanical resistance).

**Keywords:** Geopolymer composite; basalt fibers; vermiculite; mechanical properties; thermal properties.

## 1 Introduction

Research and development of high-temperature resistant materials such as geopolymers is essential to improve the safety of buildings and infrastructures in residential areas. [1] Geopolymer composites provide a more environmentally friendly and more efficient alternative to traditional materials and contribute to long-term sustainability and fire protection of buildings. [2] Composites with fibers and other fillers can be used as protective fireproof coatings or to produce boxes for storing electric batteries while they are being charged. The combination of vermiculite as an innovative solution for fire protection and basalt fibers as a guarantee of sufficient mechanical strength and durability creates a set of composite properties that make materials of this type suitable for limiting the spread of fire. [3]

The mineral vermiculite is the result of the action and chemical decomposition of rock of volcanic origin. The chemical composition of vermiculite in weight percentages (wt. %) is: $SiO_2$ 37-42; MgO 12-14; $Al_2O_3$ 10-13; $Fe_2O_3$ 5-17; $H_2O$ 8-18, and FeO 1-3. [4-6] Expanded vermiculite is a lightweight, porous material produced by heating natural vermiculite, causing it to exfoliate and expand. It has excellent insulation properties, is non-combustible, and chemically stable. In its production, the rock is heated to temperatures between 650-1000°C. The heating expands the vermiculite into large, low-mass pieces. The quality of vermiculite (the rate of expansion during production) is fundamentally affected by the presence and quantity of potassium. Expanded vermiculite has a large, chemically active surface, which makes it useful as an absorbent in some chemical processes. The main producers of vermiculite are the United States, Brazil, South Africa, Zimbabwe, Bulgaria and India. [4, 6, 7]

## 2 Experiment

Mechanical tests are aimed at comparing the strength characteristics of composite materials and their ability to withstand mechanical stress under different temperature conditions. The goal is to determine the optimal amount of vermiculite and basalt fibers. As part of the temperature tests, the thermal properties of the samples (thermal diffusivity of the composites) were analyzed. The results obtained showed the optimal composition of the composite, which meets the requirements for thermal and mechanical resistance.

The samples were made from geopolymer mixtures containing 37.5 wt.% metakaolin, 12.5 wt.% sand, and 50 wt.% alkali activator. Other ingredients added to the mixture are listed in Table 1.

Tab. 1: Geopolymer samples composition.

| Sample identification | Vermiculite (g) | Basalt fibers (g) | Aluminium powder (g) |
|---|---|---|---|
| **A** - GP *reference sample* | - | - | - |
| **B** - GP_200V_P | 200 | - | 5 |
| **C** - GP_200V | 200 | - | - |
| **D** - GP_150V_50BFs | 150 | 50 | - |
| **E** - GP_100V_100BFs | 100 | 100 | - |

Samples measuring 40×40×160 mm³ were prepared. The samples were cured for 24 hours at room temperature and then for three hours in a temperature chamber at 60 °C. Half of the samples were heated in a furnace at 800 °C, a process that should simulate thermal damage to the samples at high temperatures. After the preparation of the composite samples, the material was characterized, bending tensile strength tests, pressure tests were performed, and thermal conductivity was evaluated for the samples in the basic state and after thermal stress at 800 °C.

**Bending tensile strength**

The test was carried out according to ČSN EN 196-1 Methods of testing cement - Part 1: Determination of strength on a universal tearing machine Testometric M500-100 CT with a maximum force of 100 kN. The test sample is placed with the side wall on the cylindrical support at a defined distance. The direction of loading is perpendicular to the direction of compaction of the sample. Test parameters: diameter of cylindrical supports 10 ± 0.5 mm; distance of supports 120 ± 0.5 mm; loading speed 5 mm/s; two result datasets: $R_f$ – strength of samples without heating, $R_{f800}$ – strength of samples after 800 °C heating.

**Compressive strength**

The test was carried out according to ČSN EN 196-1 on a TESTOMETRIC M500-100 CT tearing machine. The sample is placed on the side wall between steel plates that define the size of the compressive surface of the irregular fracture. The plates are 40×40×5 mm, and during the test, the upper and lower plate fixing must be firm so that the resultant load passes through the centre of the test block. Test conditions: maximum loading force 100 kN, loading speed 5 mm/s. Dimensions of the test blocks: 40×40×80 mm³; two result datasets: $R_c$ – compressive strength of samples without heating, $R_{c800}$ – compressive strength of samples after 800 °C heating.

**Thermal diffusivity**

Thermal diffusivity describes a material's ability to compensate for temperature differences when heat is continuously dissipated by conduction in a homogeneous environment. This property is defined as the thermal conductivity ratio, the material's bulk mass product, and the specific heat capacity at constant pressure. The thermal diffusivity is determined from the natural cooling by uniform temperature at the beginning of the measurement due to thermal convection of the surrounding air environment according to Eq. (1):

$$a_i = L^2 \times \ln \frac{\frac{(t_p - t_0)}{(t_i - t_0)}}{\frac{\tau}{1}} = \frac{L^2}{\tau} \times \ln\left(\frac{(t_p - t_0)}{(t_i - t_0)}\right) \tag{1}$$

*$a_i$ - thermal diffusivity [m²/s]; L - the test body half height [m]; $t_p$ - initial surface temperature [°C]; $t_0$ - ambient temperature [°C]; $t_i$ - measured temperature at the defined time [°C]; $\tau$ - time [s].*

**Thermal conductivity**

The thermal conductivity coefficient determines the heat that must pass through a material in 1s to produce a unit temperature gradient of 1 Kelvin per square meter. It is assumed that heat propagates in only one direction. This coefficient is a characteristic property of the material. The thermal conductivity coefficient was measured using the ISOMET 2104 device, which allows direct measurement. The temperature is recorded at a defined sampling rate (number of records/min) and evaluated as a function of time.

## 3 Result and Discussion

Mechanical properties were measured for two sample sets: thermally unaffected and thermally affected at 800 °C. The measurement results were processed into boxplot graphs. The highest bending tensile strength was achieved by sample (D) (150 V + 50 BFs), the value exceeded 4 MPa (Fig. 1). The addition of 50 g of basalt fibers to the mixture resulted in a strength increase of 110 % compared to the sample without fibers (reference sample), similar to studies [8, 9]. Sample (E) with the addition of 100 g of basalt fibers achieved a similar strength value to the reference sample. Geopolymer foamed with aluminium powder, sample (B), showed 46 % and the sample with vermiculite (C) 25 % lower strength values compared to the reference sample; the results are similar to those in the publication [10] (Fig. 1, blue boxplots).

The heat-affected samples at 800 °C (Fig. 1, orange box plots) showed bending tensile strength values significantly lower than the reference sample. Only sample D came close to the value of the reference sample (2 MPa) (the measured value is 35 % lower). The highest difference in strength of the heat-treated samples is between samples A and D, the strength of sample D being 486 % higher. [8-11]

The highest compressive strength was achieved by sample D without thermal treatment (16 MPa); the fibers caused a 24 % increase in compressive strength compared to the heat-unaffected sample. The lowest compressive strength was measured for specimen B (foamed with Al powder) with an 80 % decrease (Fig. 2, blue boxplots). As in the case of the evaluation of strength results, the compressive strength values of the heat-treated samples (Fig. 2, orange boxplots) are significantly lower; thermally affected sample A (3 MPa) has a compressive strength 70 % lower than thermally unaffected sample A (13 MPa). [8, 9, 11]

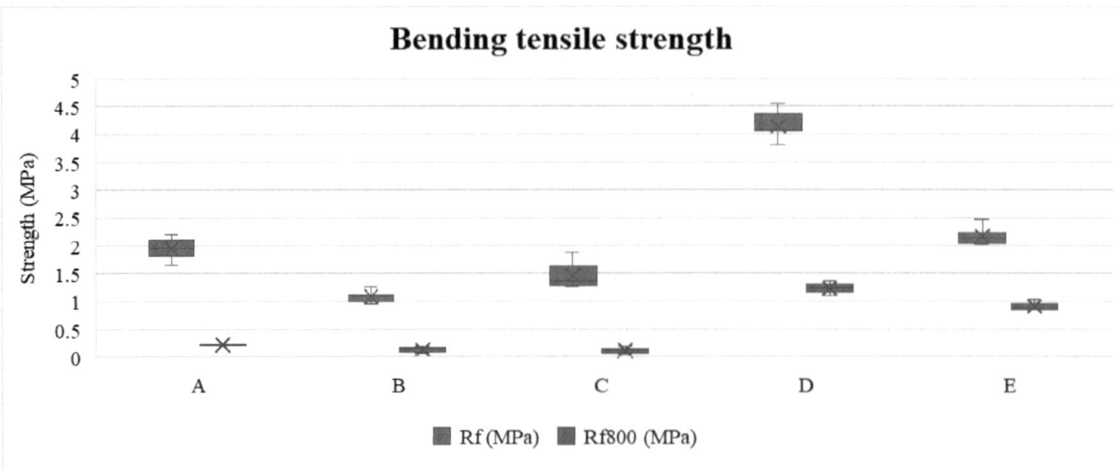


Fig. 1: Bending tensile strength.

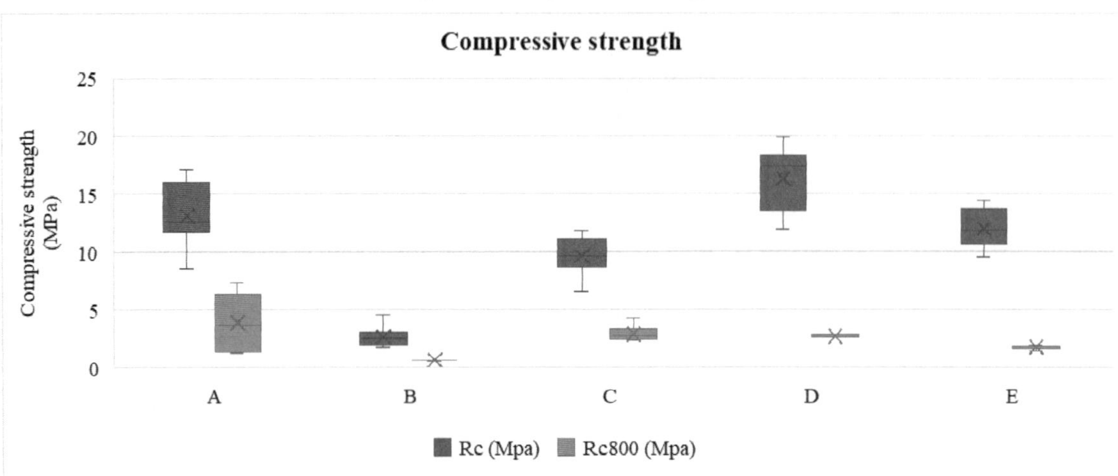


Fig. 2: Compressive strength.

The calculated thermal diffusivity values are evaluated graphically in Fig. 3. The individual samples differ in their specific values and the shape of the curve; they correspond to the typical behavior of ceramic and geopolymer materials (at lower temperatures, heat accumulation increases and heat transfer decreases, the range of values 0.4 - 1 $mm^2/s$). The highest thermal diffusivity was measured for sample B (foamed geopolymer), caused the porous structure and a low volume of heat-accumulating material. The geopolymer composites with higher vermiculite content, C (200 g) and E (150 g), followed. The lowest value was achieved by the reference sample (geopolymer without additives).

The data obtained and processed (Fig. 4) show that pure geopolymer A has a thermal conductivity of approximately 0.6 W/m·K, corresponding to the usual values for this type of material. Sample B shows the lowest thermal conductivity of ≈ 0.35 W/m·K. The material's high porosity resulted in better insulation properties. A study [10] reported that the thermal conductivity value decreases with higher vermiculite content. Samples C and E have similar values; their thermal conductivity exceeds 0.4 W/m·K and is about 30 % lower than the reference sample A. Sample D (150 G of vermiculite, 50 g of fibers) has the highest value. The thermal conductivity exceeded 1 W/m·K, indicating that a specific fiber-to-filler ratio significantly increases the material's heat transfer capacity, contradicting the study's results [10].

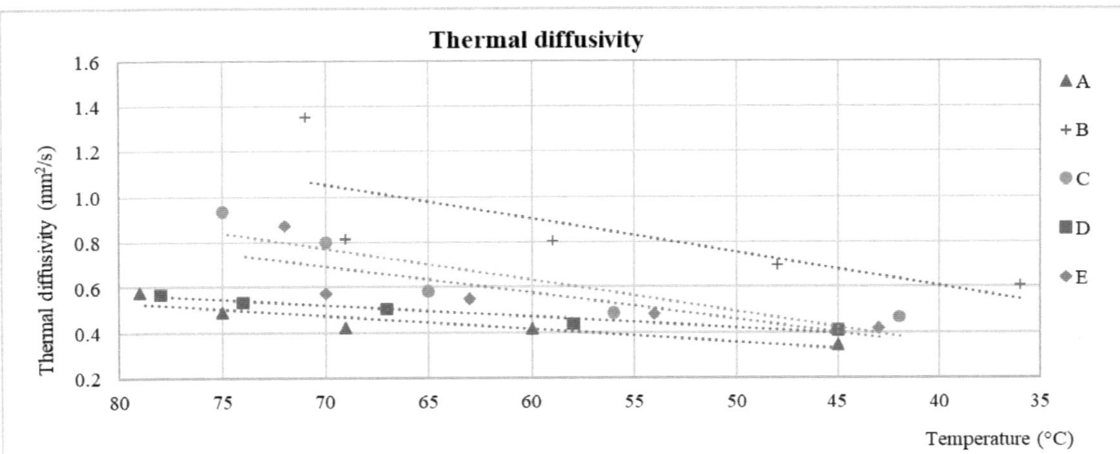


Fig. 3: Thermal diffusivity of temperature untreated samples.

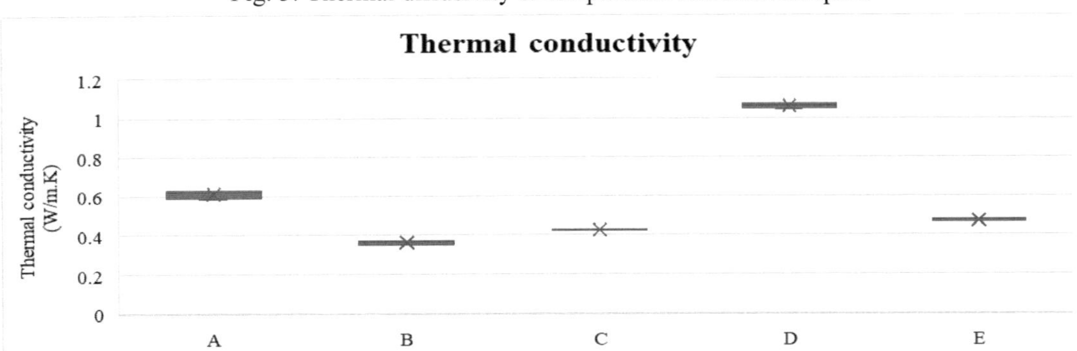


Fig. 4: Thermal conductivity of temperature-un-treated samples.

## 4 Conclusion

Basalt fibers as an additive positively affected the composite samples' mechanical properties at RT and after thermal treatment at 800 °C. On the other hand, the samples with vermiculite addition or foamed samples (foamed with Al powder) showed deteriorated mechanical properties and reached almost zero values.

The composites' durability and resistance improved for the samples with basalt fibers only at room temperature. Samples heated to 800 °C reached identical, very low values.

The thermal conductivity test confirmed that samples containing vermiculite or high porosity are suitable for thermal applications. Sample C achieved 31% better thermal conductivity values, and sample B even 40% better (compared to the reference sample).

Evaluation of the thermal insulation properties of composite materials showed that adding vermiculite or ensuring the porous structure of the geopolymer with Al powder reduces heat dissipation by 30% compared to the reference sample. The presence of a certain amount of basalt fibers leads to an increase in thermal conductivity by 60% compared to the reference sample.

The combination of basalt fibers and vermiculite may be a suitable way to optimise the geopolymer composite's mechanical and thermal properties. Further research should focus on optimising the amount and ratio of both components, which will ensure the required functional properties of the composite; a detailed analysis of the material's microstructure under various temperature and humidity conditions.

## Acknowledgement

This publication was supported by the Technical University of Liberec Student Grant Competition within the project SGS-2025-5538 and the Institutional Endowment for the Long-Term Conceptual Development of Research Institutes, as provided by the MSMT of the Czech Republic 2025.

## References

[1] J. Provis, J. Van Deventer. Geopolymers: Structures, processing, properties and industrial applications, Oxford Cambridge New Delhi, Woodhead publishing limited, 2009.

[2] L. Vickers, A. van Riessen, W. D. A. Rickard. Fire-Resistant Geopolymers: Role of Fibres and Fillers to Enhance Thermal Properties, Springer, 2014.

[3] F. Koksal, O. Gencel, M. Kaya. Combined effect of silica fume and expanded vermiculite on properties of lightweight mortars at ambient and elevated temperatures, Construction and Building Materials 88, 175-187, 2015, https://doi.org/10.1016/j.conbuildmat.

[4] A. M. Rashad, Vermiculite as a construction material - a short guide for civil engineer, Online. Construction and Building Materials, Volume 125, 2016, Pages 53-62, ISSN 0950 60 0618, https://www.sciencedirect.com/science/ar ticle/pii/S0950061816312831?via%3Dihub.

[5] P. C. d. Assis Neto, L. P. B. Sales, P. K. S. Oliveira, I. C. d. Silva, I. M. d. S. Barros, A. F. d. Nóbrega, & A. M. P. Carneiro, (2023). Expanded Vermiculite: A Short Review about Its Production, Characteristics, and Effects on the Properties of Lightweight Mortars. *Buildings*, *13*(3), 823. https://doi.org/10.3390/buildings13030823

[6] Vermiculite. Retrieved from: https://www.madehow.com/Volume-6/Vermiculite.html.

[7] Vermiculite structure, online. Available from: https://www-accessscience-com.ezproxy.tech lib.cz/content/article/a730400?implicit-login=true.

[8] M. Ziada, S. Erdem, Y. Tammam, S. Kara, R.A.G. Lezcano. The Effect of Basalt Fiber on Mechanical, Microstructural, and High-Temperature Properties of Fly Ash-Based and Basalt Powder Waste-Filled Sustainable Geopolymer Mortar. Sustainability. 2021; 13(22):12610. https://www.mdpi.com/2071-1050/13/22/12610.

[9] F. Şahin, M. Uysal, O. Canpolat, Y. Aygörmez, T. Cosgun, H. Dehghanpour. Effect of basalt fiber on metakaolin-based geopolymer mortars containing rilem, basalt and recycled waste concrete aggregates, Construction and Building Materials, Volume 301, 2021. https://www.sciencedirect.com/science/ar ticle/pii/S0950061821018730.

[10] R. Pei, L. Hua, H. Zhao, X. Wang, S. Li, Z. Wu. Experimental study on mechanical and thermal insulation properties of a geopolymer-based fireproof sandwich panel. Int J Appl Ce ram Technol. 2024; e14966. library.wiley.com/doi/10.1111/ijac.14966. [cit. 2025-01-05].

[11] M. Zaida, S. Erdem, R. A. González-Lezcano, Y. Tamman, I. Unkar. Influence of various fibers on the physico-mechanical properties of a sustainable geopolymer mortar-based on metakaolin and slag, Engineering Science and Technology, an International Journal, Volume 46, 2023, 101501, ISSN 2215-0986, https://doi.org/10.1016/j.jestch.2023.101501.

# Experimental and Numerical Investigation of Auxetic Structures Using Photoelasticity and FEM

B. Schürger [1, a], P. Frankovský[1], M. Pástor [1], P. Lengvarský[1], J. Kostka [1], P. Palička [1]

[1] *Department of Applied Mechanics and Mechanical Engineering, Faculty of Mechanical Engineering, Technical University of Košice, Letná 1/9, 042 00 Košice-Sever, Slovakia*

[a] *barbara.schurger@tuke.sk*

**Abstract:** This study examines photoelasticity as a tool for validating FEM simulations of auxetic structures. Stress distribution in planar auxetic models was analyzed under symmetrical and asymmetrical loading. Experimental measurements using PSM-1 material showed deviations of less than 5% from FEM predictions, demonstrating strong correlation. While FEM was accurate, minor discrepancies appeared in low-stress regions, influenced by model refinement and material imperfections. The experimental approach captures real-world effects, including production inconsistencies, often idealized in numerical models. These findings highlight the effectiveness of photoelasticity in stress analysis, particularly for complex auxetic geometries, and its role in refining computational models. Photoelasticity proves valuable for engineers working with FEM, even without advanced digital processing.

**Keywords:** photoelasticity; FEM; mechanical stress analysis; auxetic; experiment.

## 1 Introduction

Metamaterials are artificially engineered structures whose exceptional mechanical properties arise from their internal geometry rather than material composition [1]. Advances in additive manufacturing—such as 3D printing and stereolithography—have significantly accelerated their development [2]. A prominent subclass are auxetic structures, which exhibit a negative Poisson's ratio: they expand laterally under tension. This counterintuitive behaviour, first described by Gibson et al. [3] and later demonstrated in foams by Lakes [4], results from geometries such as re-entrant or rotating cells [5, 6] and offers improved energy absorption and stress distribution [7]. In recent years, auxetic structures have therefore become a major focus of research in mechanics and materials engineering.

Theoretical models of auxeticity were introduced by Wojciechowski [8], and the term was later formalized by Evans [9]. Auxetic structures can be isotropic or anisotropic, the latter showing directionally dependent auxetic effects [10].

Photoelasticity is a valuable method for visualizing stress distributions in these materials. Isochromatic fringe patterns reveal differences in principal stresses and have shown that auxetic designs typically display more uniform stress fields and higher strain responses than conventional structures [11]. Modern digital photoelasticity—including RGB and machine-learning-based techniques like FringeNet—has improved resolution and supports FEM validation [12].

Auxetic metamaterials are increasingly used in applications such as protective equipment [13], biomedical implants [14], and tunable mechanical systems [15]. This study validates FEM simulations of two auxetic variants using transmission photoelasticity under both symmetric and asymmetric loading. Repeated measurements show strong agreement with numerical predictions, confirming the method's value not only for model verification but also for refining boundary conditions. Although similar work has been presented by Carta et al. [16] and Köllner et al. [17], photoelastic validation remains underexplored—this work aims to extend its relevance to auxetic systems.

## 2 Materials and Methods

This study employs two auxetic models based on re-entrant cell geometries, which are among the most frequently used structures exhibiting a negative Poisson's ratio due to their characteristic inward-angled cell walls. These configurations are advantageous not only for their auxetic response but also for their manufacturability at low cost. The auxetic behavior of such cells is primarily governed by the angle θ (Fig. 1), which, according to [18], must be kept between -5° and -70° to avoid premature contact between elements under loading.

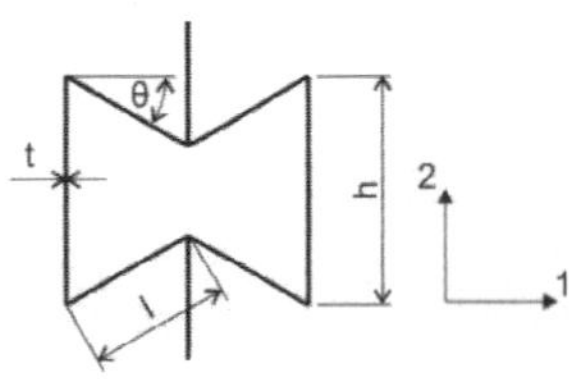


Fig. 1: Geometry of the re-entrant unit cell.

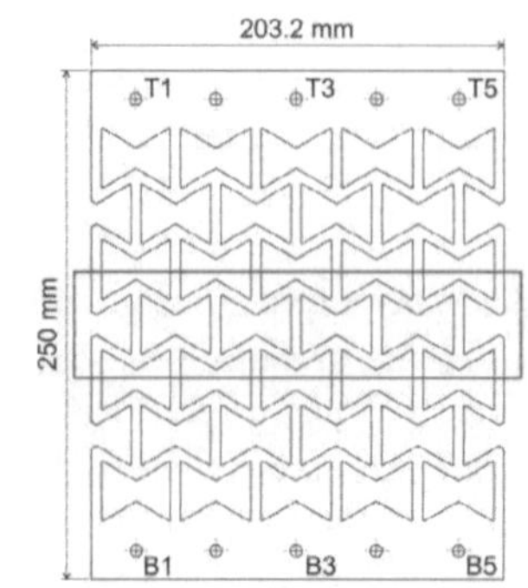


Fig. 2: Variant 1 – horizontal layout.

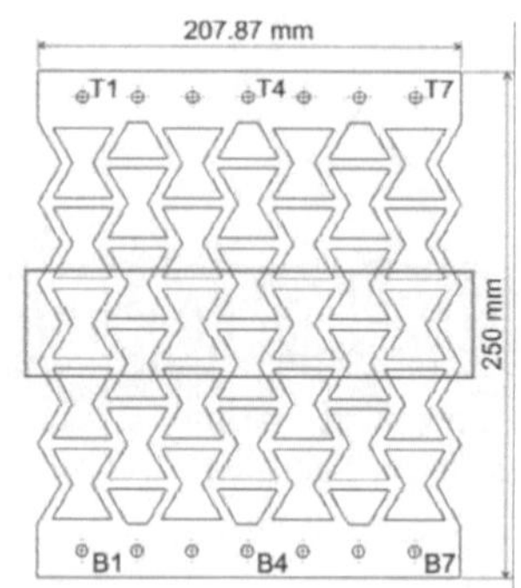


Fig. 3: Variant 2 – vertical layout.

Each unit cell is defined by its inclined wall length *l*, vertical segment *h*, and wall thickness *t* [18]. In this study, the following geometric parameters were applied: θ = -30°, h = 30 mm, l = 20 mm, and t = 5 mm.

Two physical sample variants were fabricated. Variant 1 (Fig. 2) consists of 32 full cells and 6 half cells arranged horizontally, with 5 equally spaced holes at the top (T1–T5) and bottom (B1–B5) ends for load application and fixation. Variant 2 (Fig. 3) features an identical number of cells rotated by 90°, with holes positioned according to cell distribution and labeled as Top-T and Bottom-B.

Both models were laser-cut from photoelastic material PSM-1 (Micro-Measurements, USA), a polycarbonate-based, isotropic polymer known for its excellent optical sensitivity, transparency, and mechanical stability. The material properties are listed in Tab 1.

Tab. 1: Properties of PSM-1 photoelastic sheet.

| Parameter | Value |
|---|---|
| Stress-optic constant C [MPa/fringe/mm] | 7 |
| Young's modulus E [MPa] | 2500 |
| Poisson's ratio μ [-] | 0.38 |
| Elongation [%] | 5 |
| Max temperature [°C] | 150 |
| Thickness t [mm] | 3.2 |

During fabrication by waterjet cutting, it was observed that regions with sharp interior angles were slightly rounded (radius r = 0.5 mm). To maintain accuracy, these modifications were incorporated into the final CAD models prior to simulation.

The experiment applied a maximum tensile load of 150 N, selected to match the sensitivity of the optical material. FEM simulation was used to guide measurement setup, especially in identifying regions with high stress gradients. While numerical models allow precise control of geometry and boundary conditions, they often overlook factors like manufacturing tolerances or friction. Therefore, photoelastic analysis was applied for experimental validation. This method enables full-field visualization of stress fields and helps identify discrepancies between idealized models and real structures.

Although digital methods for photoelastic evaluation—such as those based on machine learning or RGB fringe analysis—offer automation and enhanced accuracy [13], this study deliberately employed manual fringe evaluation to demonstrate the effectiveness of simpler, less resource-intensive methods. Measurement points were selected based on FEM stress maps, focusing on areas of interest (highlighted in red in Fig. 2) for direct comparison.

## 3 Numerical analysis

The finite element method (FEM) is a widely used tool for analyzing auxetic structures, particularly for evaluating their effective mechanical behavior and negative Poisson's ratio [19, 20]. In this work, the model geometry was adapted to suit experimental constraints, including the size of the polariscope and dimensions of the photoelastic sheets.

The 3D models were created in SOLIDWORKS 2024 (Dassault Systèmes, Paris, France). A mesh sensitivity analysis was performed to determine the optimal element size. Final models used tetrahedral elements with quadratic interpolation, reaching a mesh convergence where the variation in principal stress remained below 1%.

For Variant 1, the FEM mesh consisted of 388,179 nodes and 211,365 elements; Variant 2 contained 344,193 nodes and 199,863 elements. The maximum element size was set to 1 mm, with refined zones down to 0.1 mm near stress concentrators.

The structures were analyzed under static tensile loading in two modes: symmetrical and asymmetrical, as summarized in Table 2. In all cases, a total force of 150 N was applied to selected top nodes (T-points), while the bottom nodes (B-points) were fully constrained. The selected load ensures that material behavior remains within the linear elastic range of PSM-1.

Tab. 2: Boundary conditions used in FEM simulations.

| Model | Variant | Load points | Fixation points |
|---|---|---|---|
| Symmetric | 1 | T1, T3, T5 | B1, B3, B5 |
| Symmetric | 2 | T1, T4, T7 | B1, B4, B7 |
| Asymmetric | 1 | T1 | B1, B3, B5 |
| Asymmetric | 2 | T1 | B1, B4, B7 |

Stress results were evaluated using the von Mises equivalent stress criterion and are visualized in Fig. 4 and Fig. 5 for both sample variants under each loading scenario.

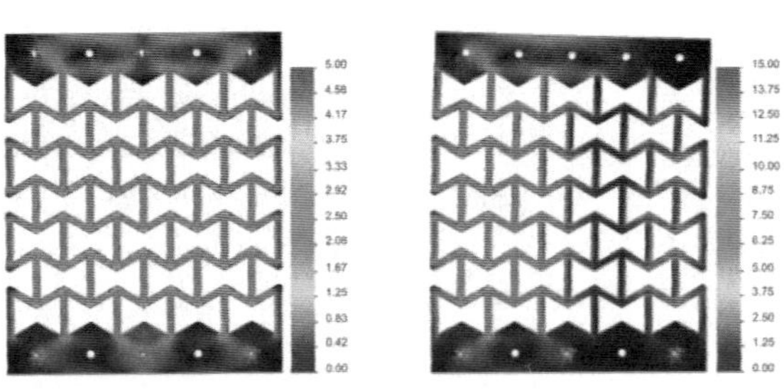

Fig. 4: Variant 1: Equivalent von Mises stress distribution (MPa), symmetrical and asymmetrical loading.

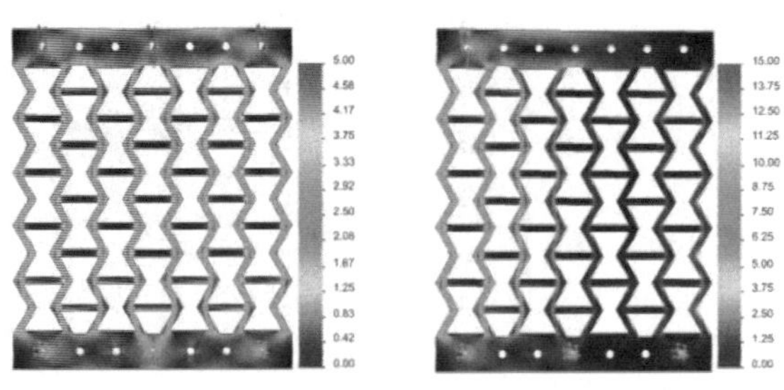

Fig. 5: Variant 2: Equivalent von Mises stress distribution (MPa), symmetrical and asymmetrical loading.

Fig. 6 illustrates the distribution of principal stresses ($\sigma_1$ and $\sigma_3$) for Variant 1 under symmetrical loading, visualized as a vector field. This representation highlights the directional changes of principal stresses, especially around areas with geometric discontinuities. In zones with uniaxial stress, vectors are aligned either parallel or perpendicular to the structural edges. These results served as the basis for selecting specific locations for experimental validation.

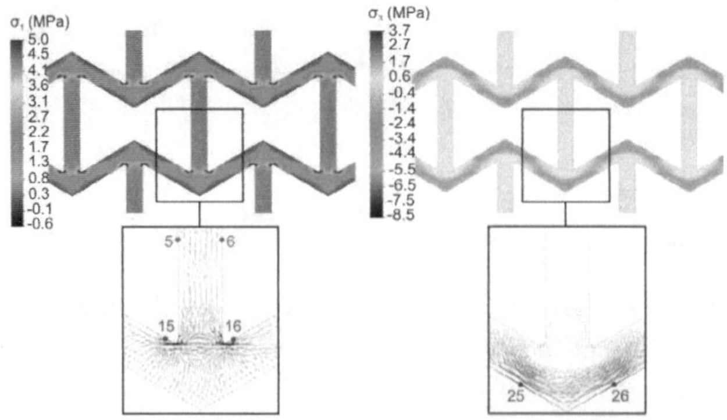


Fig. 6: Principal stress field of Variant 1: full-field vector map and detail near a stress concentrator.

For Variant 1, measurement points were chosen in the central region of the structure (Fig. 7), including:

- Vertical arms (Points 1–10),
- Areas near stress concentrators (Points 11–20),
- Regions near cell tips (Points 21–30).

As direct experimental readings within sharp concentrators are not feasible, Points 11–20 were selected in adjacent zones with high stress gradients based on FEM predictions.

Similarly, in Variant 2 (Fig. 8), points were located:

- Along the midline near concentrators (Points 1–14),
- At inner and outer tips where high stresses were observed (Points 15–30).

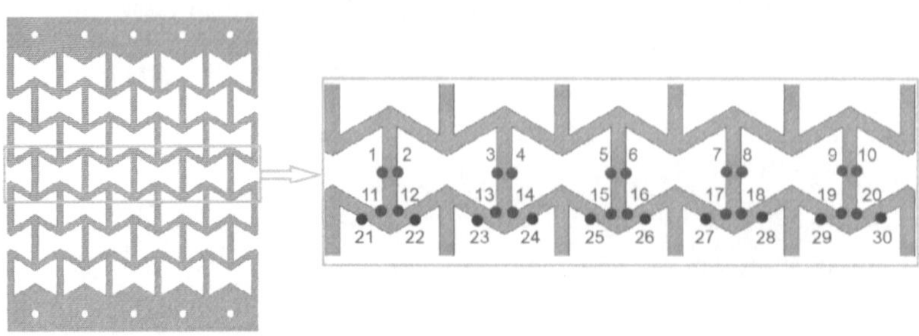


Fig. 7: Variant 1: investigated points in the central part.

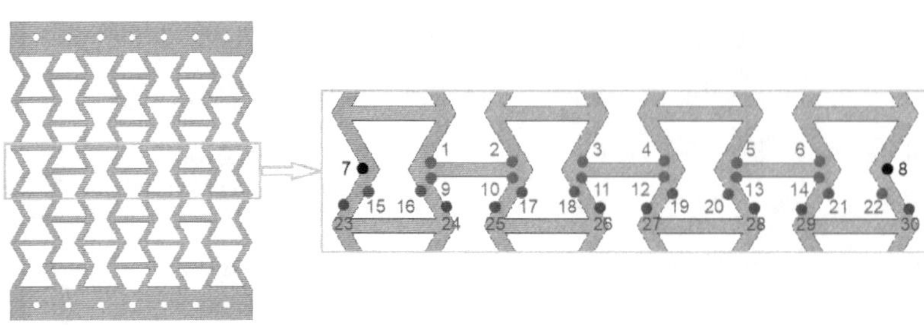


Fig. 8: Variant 2: investigated points in the central part.

For each simulated configuration, the stress values at selected points were extracted in terms of principal stresses $\sigma_1$ and $\sigma_3$, as well as the equivalent stress according to von Mises theory, which is commonly used to evaluate material strength under complex loading conditions. The relationship between the principal and equivalent stresses was used to interpret stress intensity across the structure. In particular, when one of the principal stresses is zero, the absolute value of the non-zero stress corresponds directly to the equivalent von Mises stress.

The selected locations, shown in Fig. 7 and Fig. 8, represent key structural regions, including vertical segments, areas near stress concentrators, and inner and outer cell tips. These points were used both for numerical evaluation and subsequent experimental validation. Although full graphical outputs of the stress values are not included in this section, a detailed comparison between FEM and experimental data is presented later in the Results chapter.

This approach enabled a consistent methodology for identifying appropriate points of interest and provided a reliable framework for assessing the accuracy of numerical simulations in complex auxetic geometries.

# 4 Experimental measurements

For the experimental part, transmission photoelasticity was selected as a full-field optical method for visualizing stress distribution. When a photoelastic model is subjected to mechanical loading, the difference in principal stresses produces coloured patterns known as isochromatic fringes, visible through a circular polariscope. These fringes emerge first in highly stressed areas and gradually expand toward regions of lower stress as the load increases. Their sequence—called fringe order—remains consistent throughout the loading process and does not intersect or merge.

In photoelastic analysis, the fringe order is directly proportional to the difference between the principal stresses. In regions with uniaxial stress—typically along free model edges where one principal stress is zero—the remaining stress component can be directly determined from the observed fringe order. These areas, due to their simplicity and clarity of fringe patterns, were selected as optimal points for validating numerical results.

For the experimental validation, transmission photoelasticity was employed using a circular transmission polariscope (model M060, Vishay) equipped with a null-balance compensator (model 067) with a sensitivity of 48 counts per fringe (Fig. 9). This setup enabled accurate determination of fringe orders N, which were then converted to principal stress values. At model edges, where one principal stress is typically zero, the remaining stress could be determined directly. The stress direction (tension or compression) was inferred from the compensator's orientation relative to the edge, as shown in Fig. 9.

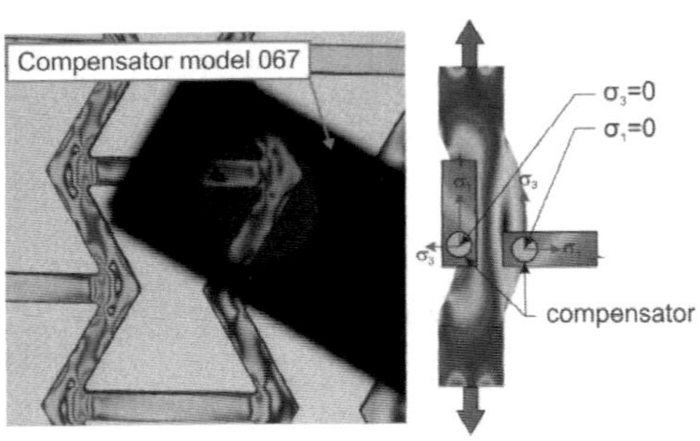


Fig. 9: Use of null-balance compensator for fringe order measurement.

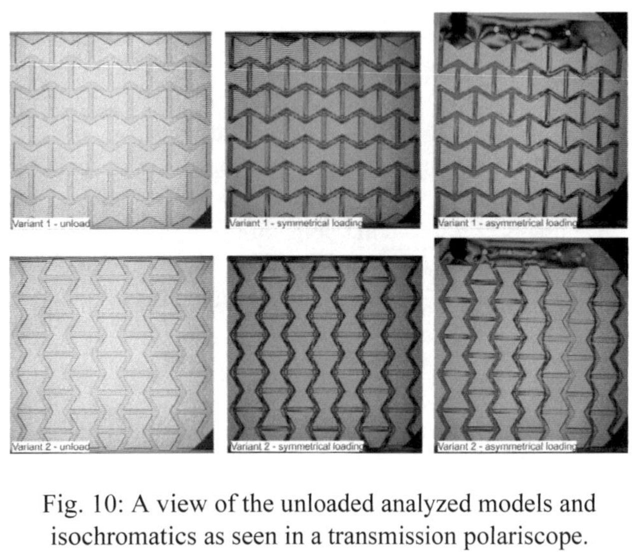

Fig. 10: A view of the unloaded analyzed models and isochromatics as seen in a transmission polariscope.

The models were cut from PSM-1 photoelastic sheets using waterjet technology, which minimizes residual stress and maintains material integrity. Each model variant was fabricated in duplicate. During loading, the applied force was measured by an RSCC force sensor (HBM) connected to a P3 data acquisition system (Micro-Measurements).

Measurement points were selected based on areas of high stress identified in the FEM simulations (Fig. 7, 8). Five repeated measurements were conducted for each model variant and loading configuration to ensure reliability. Due to the complexity of fringe interpretation in regions of extreme stress concentration, the points were chosen near the concentrators where fringe patterns remained clear and well-defined.

Fig. 10 shows the unloaded samples in the polariscope, confirming the absence of residual stresses from fabrication. The same Fig. 10 also illustrates the isochromatic fringes formed under the maximum applied load of 150 N, applied according to the conditions summarized in Table 2.

Transmission photoelasticity offers a rapid, full-field overview of the stress distribution, making it an effective method for validating numerical simulations. As expected, the highest stress levels appeared near the concentrators, visible as dense fringe regions. Differences in fringe patterns between symmetrical and asymmetrical loading were clearly observed. While fringe colour provides an initial visual cue of stress magnitude, precise measurements require operator experience or appropriate instrumentation such as the compensator used in this study.

A graphical comparison of the principal stress values obtained from FEM simulations and experimental measurements is presented in the Results section, illustrating the agreement across selected measurement points

# 5 Results and Discussion

This study demonstrates the use of transmission photoelasticity as an effective experimental method for validating numerical stress analysis performed via the finite element method (FEM). The investigated structures were auxetic planar models with a negative Poisson's ratio, a property that significantly influences the internal stress distribution. When such structures are subjected to mechanical loading, their expansion in both axial and transverse directions creates complex stress fields that are well captured by isochromatic fringes observed in photoelastic analysis.

The numerical simulations were conducted under precisely defined loading and boundary conditions, and their results were compared with experimental measurements performed on dimensionally identical samples fabricated from PSM-1 photoelastic sheets. This comparison accounts for real-world effects such as manufacturing imperfections and slight deviations in boundary constraints, which are often neglected in idealized numerical models.

The principal stress values at selected points were evaluated for both Variant 1 and Variant 2, under symmetrical and asymmetrical loading conditions. The comparison between FEM predictions and photoelastic measurements is presented in the graphs below. In most cases, the experimental values deviated by less than 5%, indicating strong agreement and confirming the reliability of the numerical model. Larger discrepancies were observed primarily in low-stress regions, where fringe visibility and accuracy of fringe order reading are reduced. Similar findings were reported in [21], highlighting that deviations greater than 5% are more likely when fringe orders are $\leq 2$. Conversely, fringe orders $\geq 2.5$ enable more reliable stress evaluation. In the graphical results presented, certain high-stress regions are not validated by experimental data, as the applied measurement method—a null-balance compensator without magnification—was limited to fringe orders up to $N = 4$. Despite this limitation, transmission photoelasticity still provided accurate stress values across a significant portion of the structure, demonstrating its capability to effectively support FEM validation even in geometrically complex auxetic models.

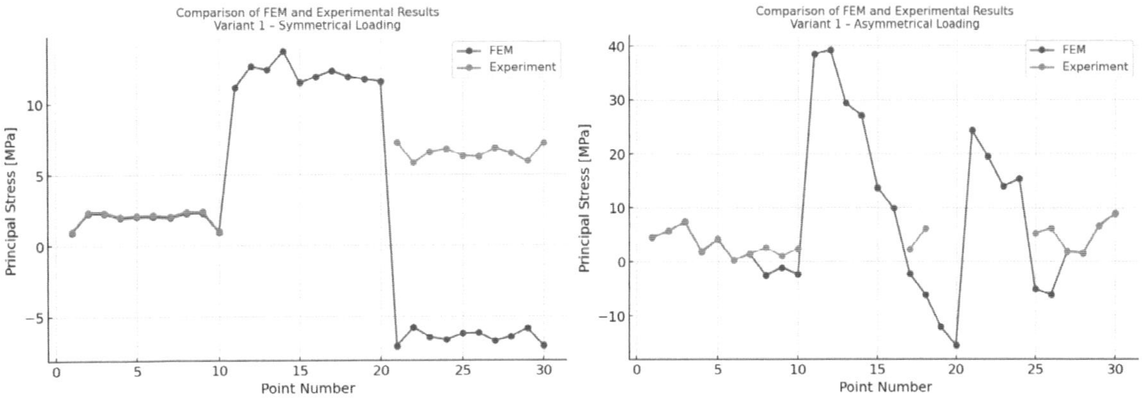


Fig. 10: Variant 1: Comparison of FEM and Experimental Results.

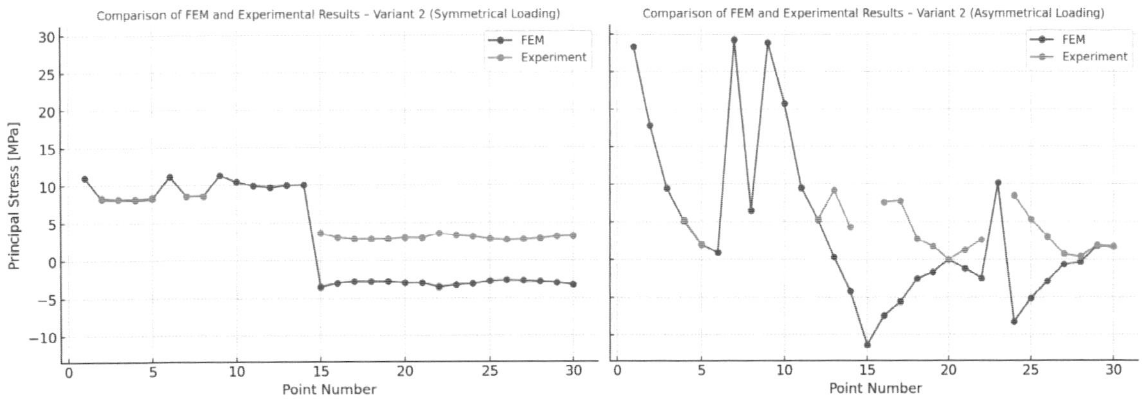


Fig. 11: Variant 2: Comparison of FEM and Experimental Results.

Importantly, the study highlights that even with basic optical tools—without the aid of modern software or high-resolution digital systems—it is possible to obtain meaningful and reliable results. This underlines the accessibility of photoelastic methods for preliminary testing or laboratories with limited resources. Nonetheless, the integration of modern optical techniques, such as digital image correlation (DIC) or full-field digital photoelasticity, could substantially enhance resolution and extend the analysis beyond edge regions to the entire model surface.

Another key finding is the importance of matching the optical sensitivity of the material to the expected stress range. In this study, the use of PSM-1 material in combination with a selected load of F = 150 N proved effective, producing well-visible fringe patterns within the measurable range. In contrast, using a material with either too low or too high sensitivity would result in unreadable patterns—either too dense (saturated) or too faint—thus compromising the quality of the experimental data.

In conclusion, the graphical comparison confirms that photoelasticity is a rapid, low-cost, and sufficiently accurate method for verifying FEM simulations, especially in studies of auxetic structures where local stress concentrations and unusual deformation patterns are expected. This method is particularly advantageous for visualizing stress distributions in a clear and intuitive way. Future research will explore the applicability of this technique to a wider range of metamaterial geometries, loading conditions, and will include hybrid approaches combining photoelasticity with digital post-processing to improve accuracy, reproducibility, and automation.


## Acknowledgement

The work was supported by the grant project KEGA 039TUKE-4/2024, VEGA 1/0152/24 and APVV-23-0461.

## References


[1] A. Álvarez-Trejo, et al., Mechanical metamaterials with topologies based on curved elements: An overview of design, additive manufacturing and mechanical properties, Mater. Design 233 (2023) 112190.

[2] C. Lu, M. Hsieh, et al., Architectural Design and Additive Manufacturing of Mechanical Metamaterials: A Review, Engineering 177 (2022) 44–63.

[3] L. J. Gibson, M. F. Ashby, G. S. Schajer, C. Robertson, The mechanics of two-dimensional cellular materials, Proc. R. Soc. Lond. A 382 (1982) 25–42.

[4] R. Lakes, Foam Structures with a Negative Poisson's Ratio, Science 235 (1987) 1038–1040.

[5] S. M. Esfarjani, A. Dadashi, M. Azadi, Topology optimization of additive-manufactured metamaterial structures: A review focused on multi-material types, Forces Mech. 7 (2022) 100100.

[6] G. Greaves, A. Greer, R. Lakes, T. Rouxel, Poisson's ratio and modern materials, Nat. Mater. 10 (2011) 823–837.

[7] C. Mercer, et al., Effects of geometry and boundary constraint on the stiffness and negative Poisson's ratio behaviour of auxetic metamaterials under quasi-static and impact loading, Int. J. Impact Eng. 169 (2022) 104315.

[8] K. W. Wojciechowski, Two-dimensional isotropic system with a negative Poisson ratio, Phys. Lett. A 137 (1989) 60–64.

[9] K. E. Evans, Auxetic polymers: a new range of materials, Endeavour 15(4) (1991) 170–174.

[10] O. Khaghani, D. Mostofinejad, S. M. Abtahi, Auxetic structures in civil engineering applications: Experimental (by 3D printing) and numerical investigation of mechanical behavior, Results Mater. 21 (2024) 100528.

[11] T. Kasimayan, K. Ramesh, Digital reflection photoelasticity using conventional reflection polariscope, Exp. Tech. 34 (2010) 45–51.

[12] A. Ajovalasit, G. Petrucci, M. Scadifi, Review of RGB photoelasticity, Opt. Las. Eng. 68 (2015) 58–73.

[13] D. Han, X. Ren, et al., Lightweight auxetic metamaterials: Design and characteristic study, Compos. Struct. 293 (2022) 115706.

[14] H. Wang, Y. Lyu, et al., A review on the mechanical metamaterials and their applications in the field of biomedical engineering, Front. Mater. 10 (2023) 1273961.

[15] C. Wu, C. Peng, T. C. Le, R. Das, P. Tran, Tunable 3D printed composite metamaterials with negative stiffness, Smart Mater. Struct. 32 (2023) 125010.

[16] G. Carta, M. Brun, A. Baldi, Design of a porous material with isotropic negative Poisson's ratio, Mech. Mater. 97 (2016) 67–75.

[17] D. Köllner et al., Advanced Estimation of Compressive Strength and Fracture Behavior in Ceramic Honeycombs by Polarimetry Measurements of Similar Epoxy Resin Honeycombs, Materials 15 (2022) 2361.

[18] X. Zhang, H. Ding, L. An, X. Wang, Numerical Investigation on Dynamic Crushing Behavior of Auxetic Honeycombs with Various Cell-Wall Angles, Adv. Mech. Eng. 7 (2015) 2.

[19] O. Khaghani, D. Mostofinejad, S. M. Abtahi, Auxetic structures in civil engineering applications: Experimental (by 3D printing) and numerical investigation of mechanical behavior, Results Mater. 21 (2024) 100528.

[20] M. Doroszko, A. Seweryn, Cyclic behaviour modelling of additively manufactured Ti-6Al-4V lattice structures, Int. J. Mech. Sci. 273 (2024) 109219.

[21] Z. Ren, H. Xie, Y. Ju, Quantification of photoelastic fringe orders using polarized light camera and continuous loading, Opt. Lasers Eng. 134 (2020) 106263.

# Tension-compression Tests of 3D-printed PETG

T. Vaňková[1,2], J. Krystek[1*]

[1] *Faculty of Applied Sciences, University of West Bohemia; Univerzitní 8, 301 00 Plzeň, Czech Republic*
[2] *Institute of Thermomechanics, Czech Academy of Sciences, Veleslavínova 11, 301 14 Plzeň, Czech Republic*
* *krystek@kme.zcu.cz*

**Abstract:** The present study focuses on the tension-compression tests of 3D-printed PETG. The anti-buckling device is designed and manufactured to perform the test. The printed specimens are exposed to tensile and subsequently compressive loading at five different temperatures. The obtained data are then processed and compared with the results of the monotonic tension and compression tests. The influence of loading and temperature on the mechanical properties of the material is analysed.

**Keywords:** experiment; PETG; tension-compression tests; temperature; anti-buckling device.

## 1 Introduction

The accessibility of polymer 3D-printing technology has a substantial impact on its utilisation in production. The current applications of 3D-printing technology are extensive, ranging from small decorative products [1] to industrial components [2] or medical implants [3]. The broad spectrum of its applicability gives rise to a multitude of demands on the material properties of the final product.

In certain instances, components undergo a transition between tensile and compressive loading. The present study aims to test this phenomenon in 3D-printed polyethylene terephthalate glycol-modified (PETG). The experimental measurements are involved in the execution of tension-compression tests at five different temperatures on samples fabricated by means of the fused deposition modelling (FDM) method. The obtained values are compared with data from tension and compression tests. The results from tension and compression tests of this material are presented in [4], where the influence of temperature on the mechanical properties of 3D-printed PETG is demonstrated.

In order to execute tension-compression tests, an anti-buckling device is required. For this purpose, a device based on a similar solution published in [5, 6] is developed and tested. Another device for similar tests is published in [7]. The disadvantage of their solution is that it does not allow the use of an extensometer to measure elongation. The device employed in this study is capable of conducting tension-compression and compression-tension tests to determine fracture ductility, Bauschinger effect, and permanent softening or hysteretic behavior [8, 9].

## 2 Experiments

The experiments are carried out on the rectangular specimens with approximate dimensions of 229 mm in length, 15 mm in width, and 2 mm in thickness (see Fig. 1). The specimens are printed horizontally from PETG filament with 100 % infill density and an infill orientation of $+45°/-45°$ on the Original Prusa i3 MK3 printer. The scheme of printing and the primary printing parameters are consistent with those outlined in the author's previous work [4]. A similar material as in the presented paper with the same infill density is of interest in [10], where the tension monotonic and cyclic tests are carried out for four different infill orientations. The mechanical properties determined for infill orientation $+45°/-45°$ are mostly worse compared to the other orientations. According to this study, the material is designated as only slightly anisotropic.

The measurements are performed using the universal testing machine Zwick/Roell Z050. A total of five different experiments are conducted. The experiments involve the execution of tension-compression tests at

Fig. 1: 3D-printed specimen.

five different temperatures ( −20 °C, 0 °C, 22 °C, 40 °C and 60 °C). The experiment is repeated to ensure the availability of data from three accepted measurements. The force-displacement dependencies are obtained from the tests. The elongation $\Delta l$ is measured using a biaxial extensometer with gage length 25 mm. It is known that during the compression loading process, there is a possibility of loss of specimen stability. Therefore, the application of an anti-buckling device is imperative to ensure the specimen's stability during loading.

The tested specimens are exposed to a quasi-static loading with a loading velocity of $v$ = 1 mm/min. Initially, the load is tensile until it reaches 80 % ultimate tensile force. Subsequent to unloading, the process transitions into a phase of compressive loading. The majority of the experiments were terminated due to the attainment of maximum displacement measurable by the used extensometer. The ultimate tensile force values are derived from the tensile tests that are presented in previous work [11].

## 2.1 Anti-buckling device

The device is designed to prevent buckling during compressive loading of the specimen. However, it is also crucial that its influence on the measured values be minimal. The anti-buckling device used in this study is composed of structural steel, a material selected for its durability and resilience. The device consists of two identical components whose geometry is illustrated in Fig. 2. The two components are interconnected by four bolts. The dimensions of the device have been meticulously selected to enable the attachment of an extensometer.

During the experimental procedure, the measured sample is positioned between the two components, and the bolts are then gradually tightened. Ensuring the appropriate degree of tightening is imperative to prevent the device from dislodging from the specimen. Simultaneously, it is crucial to avoid excessive gripping of the specimen to prevent any potential inaccuracies in the measured values. The anti-buckling device was subjected to a series of tests involving different levels of tightening, using springs and various types of washers. The resulting variants use only flat washers and the bolts are tightened manually. The anti-buckling device during the measurement process is illustrated in Fig. 3.

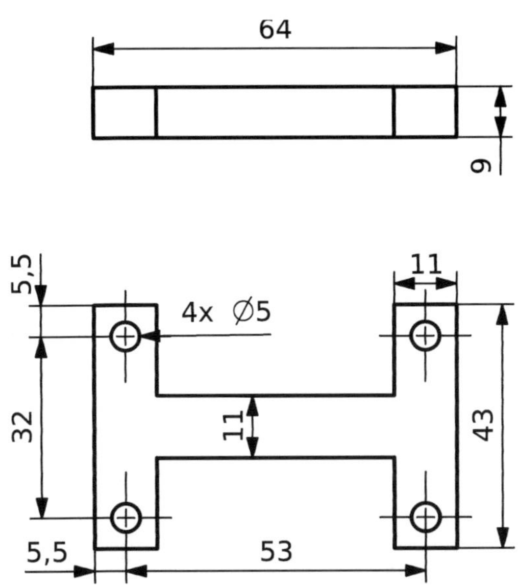


Fig. 2: Drawing of one component of anti-buckling device.

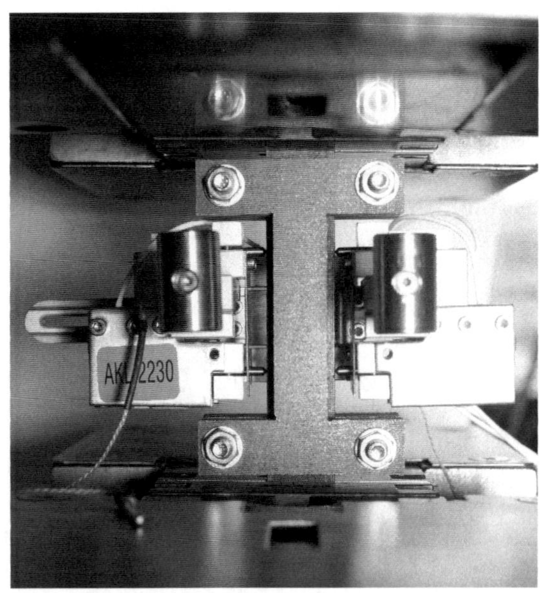

Fig. 3: Tension-compression tests with anti-buckling device and biaxial extensometer.

## 3 Data processing

The processing of the measured values (i.e., the loading force $F$ and the elongation $\Delta l$) is performed using scripts written in Python 3.5. The stress-strain dependencies are calculated from the experimentally obtained value according to the following equations:

$$\varepsilon = \ln\left(1 + \frac{\Delta l}{l_0}\right) \tag{1}$$

and

$$\sigma = \frac{F}{A_0}\left(1 + \frac{\Delta l}{l_0}\right) \tag{2}$$

where $\varepsilon$ is strain, $\sigma$ is stress, $l_0$ is initial (gage) length and $A_0$ is initial cross-section. The selection of these expressions for strain and stress is predicated on their compatibility with Abaqus software, which is planned for use in simulations of the experiments. The stress–strain curve is obtained for each test, and subsequently, the average curves are calculated for each temperature.

The Young's modulus $E$ is identified from the initial segment of the loading curve within the strain range $\varepsilon \in (0.001, 0.003)$. These values are determined from all tests as well as the maximum achieved compressive stress $\sigma_c = |\min(\sigma)|$. The mean value and its standard deviation of the obtained values at each of the measurement temperatures are presented in Tab. 1.

Tab. 1: Material characteristics obtained from tension-compression tests of 3D-printed PETG.

| Temperature [°C] | −20 | 0 | room | 40 | 60 |
|---|---|---|---|---|---|
| $E^{\mathrm{mean}}$ [GPa] | 1.97 | 1.92 | 1.91 | 1.79 | 1.54 |
| $E^{\mathrm{std}}$ [GPa] | 0.02 | 0.03 | 0.01 | 0.01 | 0.10 |
| $\sigma_c^{\mathrm{mean}}$ [MPa] | 40.7 | 42.7 | 41.3 | 38.3 | 30.4 |
| $\sigma_c^{\mathrm{std}}$ [MPa] | 0.9 | 4.5 | 0.4 | 0.5 | 1.1 |

## 4 Results

The obtained results from tension-compression tests are compared with the results from tension and compression tests. The examples of comparison of the determined stress-strain curves from tests at −20 °C and 60 °C are shown in Fig. 4. It is evident, that the damage and permanent strain caused by the given tensile load are minimal. The maximum damage and permanent strain are observed at the temperature of 60 °C.

The values of Young's modulus $E$ and the maximum achieved compressive stress $\sigma_c$ obtained from the tension-compression tests are compared with the values from tensile and compression tests. The comparison of these values depending on temperature is illustrated in Fig. 5. The values $E$ and $\sigma_c$ from these tests are analogous to the values from tensile and compression tests.

The most significant discrepancy in Young's modulus $E$ is at 0 °C. At this temperature, the two obtained values are proximate to the upper limit of the interval ($E^{\mathrm{mean}} \pm E^{\mathrm{std}}$), and one value is adjacent to the lower limit. The maximum achieved compressive stress $\sigma_c$ from tension-compression tests at temperatures of −20 °C and 0 °C appears to be less than the values obtained from compression tests. Nevertheless, the discrepancy is negligible, and additionally the deviation from the tension-compression tests at 0 °C is large. The values of $\sigma_c$ at all temperatures are influenced by the termination method used in these tests. However, the analogous of $\sigma_c$ from the tension-compression and compression tests can be attributed to the presence of only minor damage and permanent strain.

Depending on the temperature, the values of Young's modulus $E$ and the maximum achieved compressive stress $\sigma_c$ mainly decrease with increasing temperature. The influence of temperature on the tested material is observed for the stress-strain curves from tension-compression tests (see Fig. 6).

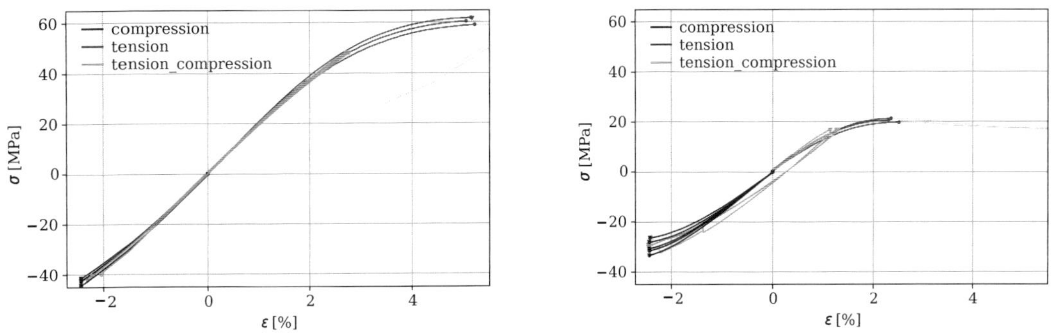


(a) Tests at −20 °C

(b) Tests at 60 °C

Fig. 4: Comparing stress–strain curves from tension, compression and tension-compression tests.

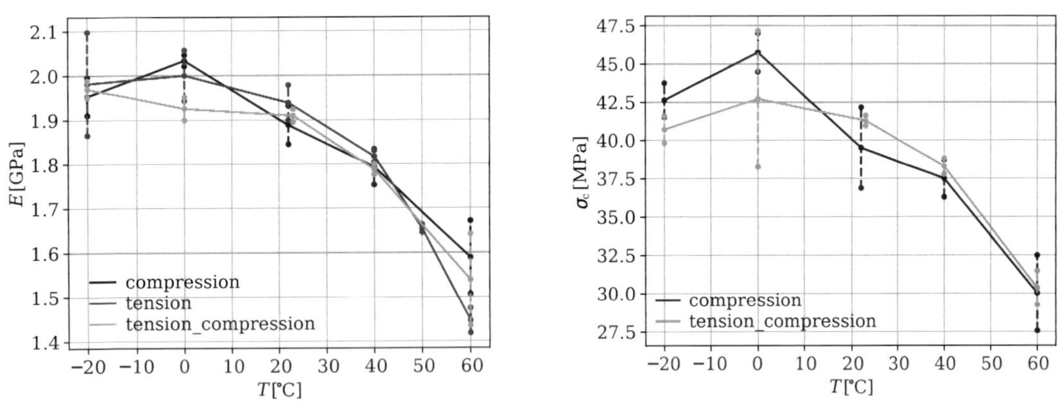


(a) Young's modulus

(b) Maximum compressive stress

Fig. 5: Comparing values from tension, compression and tension-compression tests.

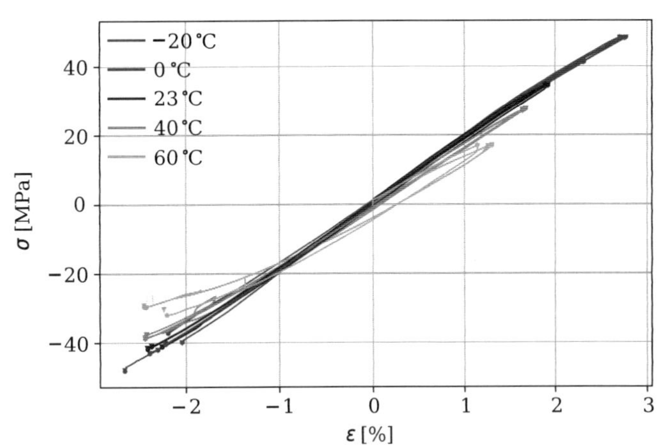


Fig. 6: Influence of temperature on stress–strain curves from tension-compression tests.

## 5 Conclusion

The tension-compression tests of 3D-printed PETG are carried out using an anti-buckling device that has been designed and manufactured for this purpose, with the objective of preventing the loss of stability. The influence of this device on Young's modulus is negligible. The permanent strain obtained after tensile loading is minimal. The maximum value of permanent strain is recorded as 0.25 % at the temperature of 60 °C.

The anti-buckling device could be used in an additional test, which might yield intriguing insights regarding the material. For instance, the compression-tension tests are employed to compare the results with those from this study and to ascertain the damage or cyclic tests with alternating loads could be used to obtain the hysteretic behavior.

## References


[1] All3DP, 40 beautiful home decor 3D printing ideas (2024), Retrieved from: https://all3dp.com/2/3d-printing-ideas-home-decor-models/

[2] T. Sathish,M. Vijayakumar,A. K. Ayyangar,Design and fabrication of industrial components using 3D printing, Materials Today: Proceedings 5 (2018) 14489–14498, doi: 10.1016/j.matpr.2018.03.036.

[3] A. M. E. Arefin et al., Polymer 3D Printing Review: Materials, Process, and Design Strategies for Medical Applications, Polymers 13 (2021) 1499, doi: 10.3390/polym13091499.

[4] T. Vaňková et al., Identification of temperature-dependent parameters for an elastic-plastic-damage material model of 3D-printed PETG, Polymer 284 (2023) 126295, doi: 10.1016/j.polymer.2023.126295.

[5] H. Liu et al., Fatigue modeling for carbon/epoxy unidirectional composites under various stress ratios considering size effects, International Journal of Fatigue 120 (2019) 184–200, doi: 10.1016/j.ijfatigue.2018.11.009.

[6] M. Stoudt, L. Levine, L. Ma. Designing a Uniaxial Tension/Compression Test for Springback Analysis in High-Strength Steel Sheets. Experimental Mechanics 57 (2016) 155–163, doi: 10.1007/s11340-016-0202-x.

[7] R.K. Boger et al., Continuous, large strain, tension/compression testing of sheet material, International Journal of Plasticity 21 (2005) 2319–2343, doi: 10.1016/j.ijplas.2004.12.002.

[8] X. Wei, Y. Liu, X. Zhang, Effect of non-linear tension-compression loading reversal on the hardening behavior and initiation fracture strain of a cold-rolled TRIP780 steel sheet, Materials Today Communications 30 (2022) 103076,doi: 10.1016/j.mtcomm.2021.103076.

[9] X. Zhou et al., Experimental study of shape memory alloy plates under cyclic tension–compression loading scenarios, Engineering Structures 297 (2023) 117002, doi: 10.1016/j.engstruct.2023.117002.

[10] G. Dolzyk, S. Jung, Tensile and Fatigue Analysis of 3D-Printed Polyethylene Terephthalate Glycol, Journal of Failure Analysis and Prevention 19 (2019), doi: 10.1007/s11668-019-00631-z.

[11] T. Vaňková et al., Tensile and Compressive Properties of 3D-printed PETG Depending on Temperature, in proc.: EAN 2022 - Proceedings of the 60th Annual Conference on Experimental Stress Analysis (2022), ed. K. Doubrava et al.,Czech Technical University, Prague 407 – 412.

# Using a friction sensor in experimental stress analysis

J. Václavík [1,a], R. Musilová [1], V. Yanovych [1]

[1] *Research and Testing Institute Plzeň; Tylova 1581/46, 30100 Plzeň, Czech Republic*
[a] *vaclavik@vzuplzen.cz*

**Abstract:** With the strain checker the strain is measured using frictional force working on the contact surface of the frictional strain gauge by pressing the gauge against the structure with magnetic force. The aim of this work is to determine the stability of the sensor during a long-term test, the possibility of measuring stress amplitudes during a cyclic test, and to show the use of this sensor in measuring structural stress during a dynamic test of a towing device.

**Keywords:** friction sensor; strain gauge; mechanical coupling device.

## 1 Introduction

Strain is easily measured by directly attaching the friction strain checker FGMH-2A series to a position of a steel bridge, crane, or any other structure where it is necessary to measure the strain. Positions of measurement can be moved, and the measurement can be repeated easily. Since adhesives are not required, strain measurements require no removal of paint, polishing, degreasing, adhesive bonding or curing. Because of this, no subsequent repainting is required after the measurement.

Additionally, stress distribution and the direction of principal stress can be obtained by simultaneous multi-point measurement using multiple strain checker. The unit itself has a gauge length of 3 mm. The gauge factor is approximately 2, while gauge resistance is 120 Ω. The completion to the whole bridge is made inside the checker.

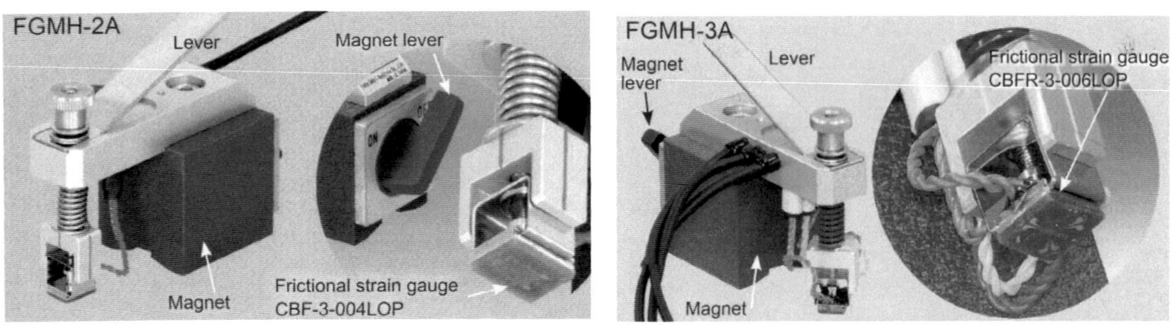


Fig. 1: Available product versions, with single strain gauge (left) and strain gauge rosette (b) (source: TML).

The introducing of this sensor was first announced in [1], some articles deal with its use and additions to the measuring system [2], [3]. Here described sensor is suitable only for stress measurement in one axis. For long-term quasi-static tests, we have developed the easy measuring unit based on Arduino MKR microcontroller and the A/D chip HX711 with the display and recording to the SD card (Fig. 2).

## 2 Basic tests with frictional sensor

The stability of the friction sensor and its ability to transfer strains were tested on a Schenck 250 kN hydraulic testing machine on a rectangular sample with a cross section of 50x4 mm (Fig. 3). The test was verified by measurements using strain gauges.

Static, sinusoidal dynamic and transition tests were performed with the sample. Strain time history of long-term test of resistance to sudden change of the loading force is presented in Fig. 4. Stepwise loading and unloading of the sample were compared with the output from strain gauges. The hysteresis loop of the sensor strain including the percentual deviation from the actual strain is given in Fig. 4b. Parallel measurements using strain gauges (SG) showed a mutual deviation of approximately 3 %.

Fig. 2: Measuring unit for strain gauge checker.

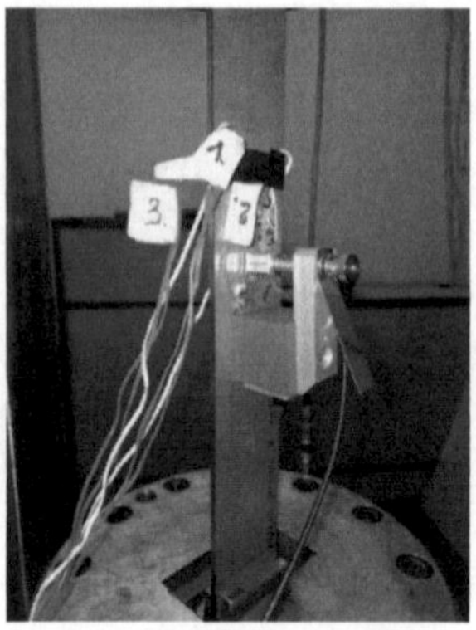

Fig. 3: Comparative tests on hydraulic frame with strain gauges.

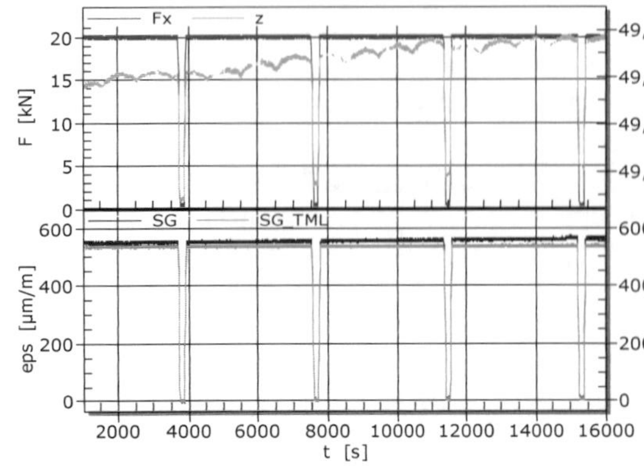


a)

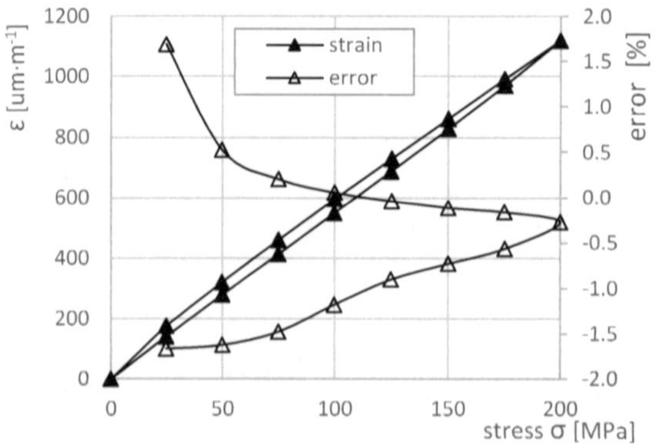


b)

Fig. 4: Visualization of performed tensile tests - long-term tests of resistance to sudden change of the load (a) and the deviation from the set of values, obtained during stepwise tensile loading (b).

## 3 Towing bracket test

Friction sensor was used to measure the stress of the bus structure (Fig. 5a) during a towing bracket test according to UN ECE R55. In our case, bracket with ball and ball support, fastened using class A50 bolt was investigated (Fig. 9b). The test sample comprises the coupling ball, the ball neck and the mountings necessary for attaching the assembly to the vehicle. With coupling devices from steel, strength was verified by an endurance test with sinusoidal load $F_a$ = 15,5 kN and frequency f = 2.5 Hz up to N = $2 \cdot 10^{6}$ stress cycles, no cracks or fractures shall occur at these test conditions. The testing force is a function of towing and towed masses. The direction of the angle of test shall be determined by the vertical relationship between a horizontal reference line passing through the center of the ball and a horizontal line passing through the fixing point of the coupling device. In our case the angle was -15°.

The FEM model of the rear bus structure was created and the stress response to maximum force amplitude was calculated (Fig. 6). The strain measurements with the friction sensor were made during dynamic load of f = 2,5 Hz which was very good in comparison with strain gauges (Fig. 7a). Stress response along the three lines on the upper and side part of the bracket, measured using strain checker from strain amplitudes, is given in Fig. 7b. The measured stress was under the expected fatigue limit, which was confirmed by testing – no damage of the bracket appeared.

a) Rear part of the bus (red arrow indicates the location of the crack)

b) Detail view of the towing bracket and applied strain checker

Fig. 5: Towing bracket test, assembled on the rear part of the simplified bus frame.

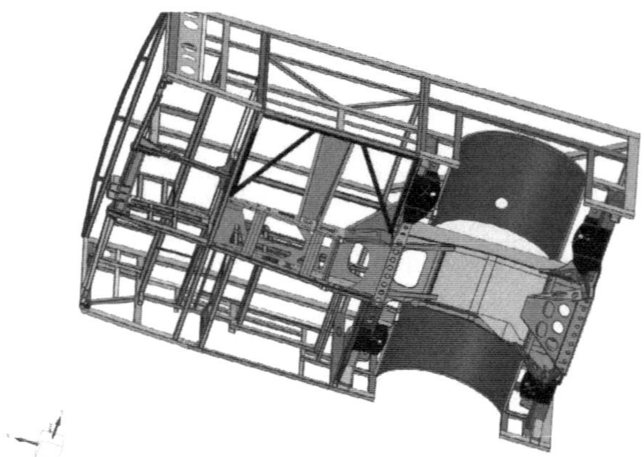

a)

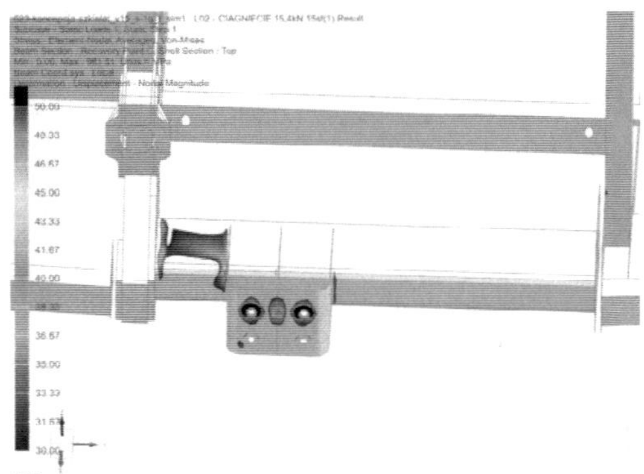

b)

Fig. 6: FEM model of the simplified testing frame (a) and calculated HMH stress field to the applied force.

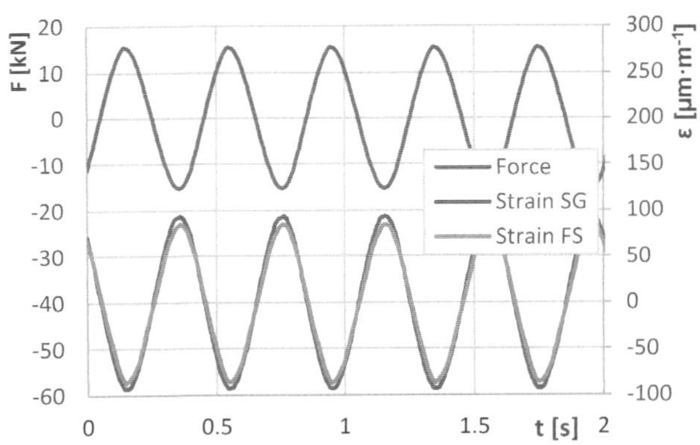


a) Response from the FS sensor to dynamic load

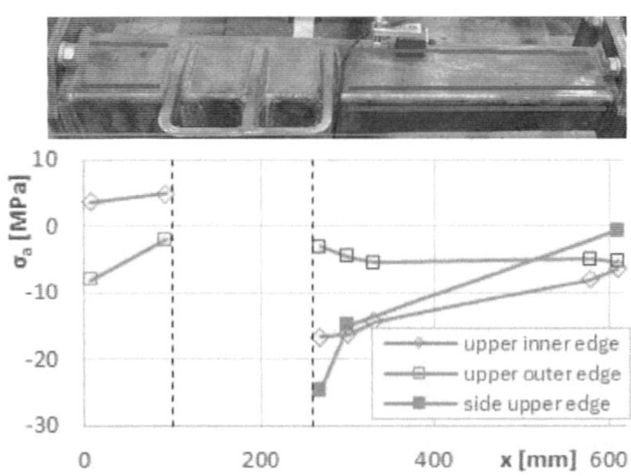


b) Stress distribution along the towing bracket

Fig. 7: Strain measurement results, calculated from the stress ranges, measured during the dynamic load.

However, high stress was measured on the bus structure at the weld, far from the tested bracket (Fig.8a), which was over the fatigue limit for N = 2E6 cycles for expected S-N curve FAT= 60MPa. A crack has indeed appeared here. Based on this fact, the new S-N curve for this joint could be created by shifting the proposed one to lower stresses. New lower FAT = 48 MPa was found, see Fig. 8b.

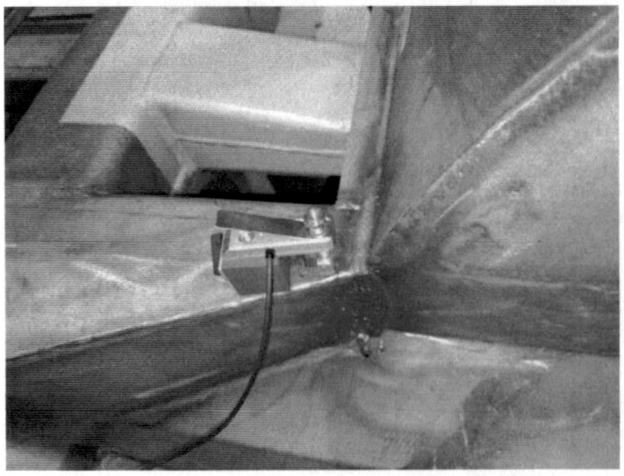

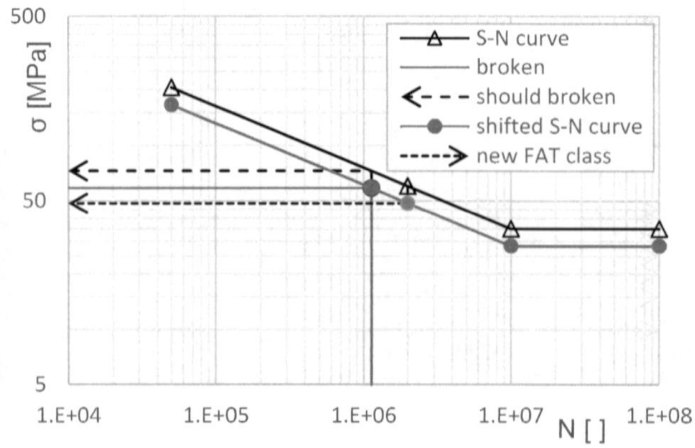


a) b)

Fig. 8 High stress measured on longitudinal beam above the structure attachment point (a), see red arrow in Fig. 5a and derivation of the new S-N curve for this joint (b).

## 4 Conclusion

The friction strain sensor is suitable for the determination of most stressed parts on the structure even during dynamic tests up to units of Hz. Multipoint strain measurement is very useful for subsequent fatigue analzsis of the structure.

Attention must be paid to the preparation of the measured surface, otherwise there is a risk of damage to the measuring grid of the strain gauge.

## Acknowledgement

The article originated in the framework of institutional support for the long-term conceptual development of the research organization.

## References

[1] J.L.Wiliamson, J.A.Coutts, An assessment of the frictional type strain gauge, Strain, January 1969, p. 15.
[2] J.Hongki at al, Develoment of high-sensitivity wireless strain sensor for structural health monitoring, Smart Structures and Systems, Vol. 11, No. 5 (2013) 477-496, DOI: http://dx.doi.org/10.12989/sss.2013.11.5.477
[3] E.Robbins et al, Development of a low-cost efficient wireless intelligent sensor for strain measurements (LEWIS-S), Meas. Sci. Technol. 32 (2021) 095109 (11pp), doi.org/10.1088/1361-6501/abe339.
[4] Uniform provisions concerning the approval of mechanical coupling components of combinations of rolling vehicle sets, Directive R55, October 2022.

# Index of authors